Sudden Loss of Cochlear and Vestibular Function

Volume Editors
M. Hoke, Münster and *M. E. Wigand,* Erlangen

88 figures and 25 tables, 1981

S. Karger · Basel · München · Paris · London · New York · Sydney

Advances in Oto-Rhino-Laryngology

National Library of Medicine, Cataloging in Publication
    Sudden loss of cochlear and vestibular function /
    volume editor, M. Hoke. – Basel; New York: Karger, 1981.
    (Advances in oto-rhino-laryngology; v. 27)
    Symposium of the Arbeitsgemeinschaft Deutscher Audiologen and Neurootologen
    held at Erlangen, West Germany, in March 1980.
    1. Cochlear – physiopathology – congresses 2. Deafness, Sudden – etiology – congresses
    W1 AD701 v. 27 [WV 270 S943 1980]
    ISBN 3-8055-2630-X

Drug Dosage
    The author and publisher have exerted every effort to ensure that drug selection and dosage
    set forth in this text are in accord with current recommendations and practice at the time of
    publication. However, in view of ongoing research, changes in government regulations, and the
    constant flow of information relating to drug therapy and drug reactions, the reader is urged
    to check the package insert for each drug for any change in indications and dosage and for
    added warnings and precautions. This is particularly important when the recommended agent
    is a new and/or infrequently employed drug.

Sudden Loss of Cochlear and Vestibular Function

Advances in Oto-Rhino-Laryngology

Vol. 27

Series Editor
C. R. Pfaltz, Basel

S. Karger · Basel · München · Paris · London · New York · Sydney

Contents

Preface

Generally speaking, a workshop is an institution which provides the means of presenting brand-new findings and preliminary data; it is the place where rough conceptions and tentative hypotheses are put forward and discussed, are rejected or supported. Thus, the publication of the proceedings of a workshop is not only an unusual procedure, but highly undesirable because the printed word is hardly compatible with improvisation and incompleteness. Incomplete to a high degree is also our knowledge of two otologic entities, the sudden hearing loss and the acute vestibular neuropathy. In spite of their frequent confrontation with patients, both the otologist and the audio-neurootologist feel urged by their ignorance concerning etiology and empirical therapy. Only one fact has emerged from the darkness: These are not two homogeneous diseases but are symptoms of totally different origin. That may be vascular disturbances, viral infections, otosclerosis or the acoustic neuroma. Accordingly, very different forms of treatment are required.

In this situation, it was the aim of the 'Arbeitsgemeinschaft Deutscher Audiologen und Neurootologen' to bring related disciplines into cooperation in order to refer to their latest knowledge of diagnostical techniques and pharmacotherapy, and the editors are grateful to those who have communicated open-minded with an adjacent specialty. It was, of course, impossible to treat all aspects comprehensively. But we ended up with quite a view remarkable contributions which was the stimulus to publish the collected contributions on the general theme. Despite some basic and more comprehensive articles, this volume still exhibits a touch of workshop character because all pertinent free papers which were presented have also been included.

I am hopeful that the findings published here may promote the understanding of disorders of cochlear function, may serve to restrict the use of the term 'sudden hearing loss' unequivocally to cochlear events, and may stimulate further progress in clinical and basic research.

The editors

Fig. 1. Inferior anterior cerebellar artery with histochemical demonstration of adrenergic fibre plexus around the vessel.

habenula to form the arcades of the terminal plexus. In combined preparations of ink perfusion of the blood vessels and histochemical demonstration of adrenergic fibres the separate course of blood vessels and adrenergic fibres is unquestionably shown in all turns (fig. 2). Only occasionally a fibre is seen to follow a blood vessel over a longer distance, which does not allow the conclusion that this is a perivascular innervation. The adrenergic fibres in the osseous spiral lamina are most conspicuous in the apical turn, where they can be followed through the entire length of the osseous spiral lamina. This corresponds possibly to the special autonomic innervation of the cochlear apex described by *Palumbi* [1950].

There is, however, no adrenergic innervation in the other parts of the membraneous labyrinth, especially not in the spiral ligament and the stria vascularis. This is consistent with the electromicroscopic (EM) observations where no nervous structures are found in the spiral ligament or in the stria. Sometimes, however, some yellow granules can be seen in such preparations within the tissue of the spiral ligament without any connection to nerve fibres. We have not yet determined which substance produces this fluorescence. It certainly, however, is not noradrenaline, which on the basis of the fluorescence colour could be excluded by different

Also in the vestibular labyrinth a blood vessel-independent adrenergic nerve fibre network is found below the sensory epithelia as clearly seen in the macula utriculi [*Spoendlin and Lichtensteiger,* 1966].

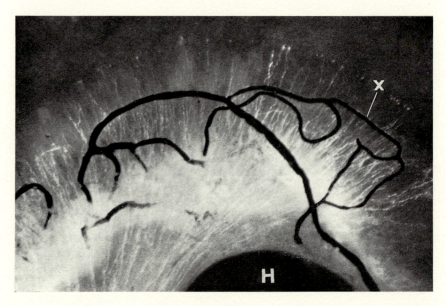

Fig. 2. Lamina spiralis ossea of the apical end of a cat cochlea with histochemical demonstration of adrenergic nerve fibres, which rund essentially radially among the other nerve fibres in the osseous spiral lamina. The blood vessels are filled with Indian ink (X) showing clearly that the adrenergic nerve fibres run independent of the blood vessels. H = Helicotrema.

filter combinations. It might be a substance related to dopamine such as melanin, which is regularly present in the stria vascularis. In any case a perivascular innervation and therefore a direct nervous influence on the blood vessels of stria and spiral ligament does not exist. Whether an indirect neurohumoral effect on these blood vessels is possible remains to be clarified.

Relationships of the Adrenergic Neurons

In EM preparations there are relatively few unmyelinated nerve fibres in the lamina spiralis ossea, which present the morphological criteria of adrenergic elements. They are of small size and have varicose enlargements filled with numerous dense core vesicles of diameters between 600

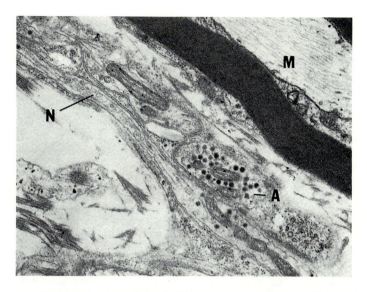

Fig. 3. Portion of the distal part of the osseous spiral lamina with a myelinated large cochlear nerve fibre (M) and some unmyelinated small adrenergic nerve fibres (N) with enlargements, filled with numerous dense core vesicles (A). Situation after administration of 5-hydroxidopamine.

and 1,200 Å. Such accumulations of dense core vesicles are typical for the adrenergic system (fig. 3). They are especially enhanced after administration of 5-hydroxidopamine, a false transmitter, which is selectively accumulated in the dense core vesicles of the adrenergic fibres [*Tranzer and Thoenen,* 1967]. Real nerve endings, as those described by *Arnold* [1974], did not show in our preparations. The adrenergic system forms rather a terminal plexus with noradrenaline stores than definite nerve endings. The transmitter is liberated from the stores upon adequate stimulation and acts in a neurohumoral way on the effector organ.

In the osseous spiral lamina the adrenergic nerve fibres rund predominantly between the myelinated fibres of the cochlear neurons without any closer relations to them. In the tympanic lip they run usually in a certain distance from the blood vessels, the wall of which consists only of endothelium and some pericytes. These adrenergic fibres and their noradrenaline stores are especially pronounced after administration of 6-hydroxidopamine, which accumulates very fast in the adrenergic neurons and finally destroys them selectively (fig. 4) [*Tranzer and Richard,* 1971].

Origin and Pathways of Adrenergic Neurons to the Inner Ear

The cells of origin of the sympathetic nervous system are situated in the upper thoracal medulla. The preganglionic fibres leave the CNS with the anterior roots and reach the sympathetic chain through the white rami communicantes. From there the postganglionic fibres join the blood vessels or segmental nerves via grey rami communicantes. They reach the periphery either with these blood vessels or nerves. Principally the same organization is found in the cervical sympathetic system where the postganglionic adrenergic fibres reach the peripheral destination together with the blood vessels, cervical and caudal brain nerves. Section of the preganglionic fibres induces usually no degeneration of the postganglionic fibres, whereas section of the postganglionic fibres or extirpation of the sympathetic ganglia causes a complete degeneration of the corresponding peripheral adrenergic fibre systems.

Extirpation of the superior cervical ganglion causes a complete disappearance of the adrenergic innervation of the cochlea on the operated side, whereas the innervation of the basilar artery complex including the labyrinthine arteries remains intact and symmetrical. The transection of the cervical sympathetic trunc below the superior cervical ganglion has no effect on the inner ear adrenergic innervation. This clearly indicates that these adrenergic fibres originate in the superior cervical ganglion [*Spoendlin and Lichtensteiger, 1967*].

In the cat, where the internal carotid artery is absent or only rudimentary, the four main branches of the superior cervical ganglion lead directly to the base of the scull, their transection has the same effect as the extirpation of the superior cervical ganglion. Destruction of the minor branches to the common carotid artery leaves the inner ear unchanged.

In order to follow the course of the main sympathetic branches at the base of a scull, careful anatomical dissections were performed in several animals. Stained with osmic acid the unmyelinated autonomic nerves are brown and the myelinated somatic nerves black. Almost all cranial branches from the superior cervical ganglion enter the bulla and the middle ear to form, together with a somatic branch from the glossopharyngeal nerve, a pronounced tympanic plexus. One or two of its ramifications anastomose with the greater superficial petrosal nerve. All others leave the middle ear anteriorly. One branch from the superior cervical ganglion, however, is found to run posteriorly without entering the middle ear. It is connected with the auricular branch of the vagal nerve which is fairly large in most

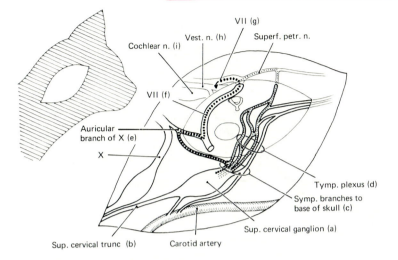

Fig. 6. Schematic representation of the pathways and connections of the sympathetic branches from the superior cervical ganglion. The sympathetic pathways are represented by interrupted lines, which essentially lead through the auricular branch of X and to the tympanic plexus to be connected either directly to the geniculate ganglion or to the superficial petrosal nerve.

mammals. An anastomosis between the auricular branch of the vagal nerve and the facial nerve is present in the cat and other mammals as well as in man.

According to these anatomical relationships the sympathetic nerves from the superior cervical ganglion can use two ways to reach the inner ear: (1) either via the tympanic plexus–anastomosis with the superficial petrosal nerve-facial nerve-internal acoustic meatus, or (2) via the auricular branch of the vagal nerve – anastomosis with the facial nerve-facial nerve-internal acoustic meatus (fig. 6). In the internal acoustic meatus the vestibulofacial and the vestibulocochlear anastomosis might serve as further pathways for the adrenergic fibres.

Transection of the tympanic plexus together with the auricular branch of X leads to a complete disappearance of the adrenergic fibres in the cochlea. Interruption of the tympanic plexus alone, however, is followed only by a partial reduction of the adrenergic innervation of the inner ear. The section of the auricular branch of X has usual a greater effect in the cat and leads to almost complete disappearance of the adrenergic innervation of the inner ear. The same occurs after section of the facial nerve

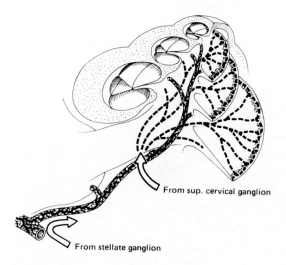

Fig. 7. Schematic representation of the double adrenergic innervation of the cochlea. One system originates in the superior cervical ganglion and is independent from blood vessels and the other originates from the stellate ganglion and is strictly perivascular.

in the oval window area. A complete loss of adrenergic fibres is also observed when the facial nerve and the vestibulofacial anastomosis is sectioned in the inner acoustic meatus. After such lesions the degeneration of the adrenergic innervation of the inner ear is always strictly ipsilateral. In spite of a complete disappearance of the blood vessel independent adrenergic fibres in the cochlea of one side the perivascular plexus around vertebral, basilar, inferior anterior cerebellar and labyrinthine artery remain always symmetrically present on the operated and unoperated sides.

The perivascular innervation of the vertebral basilar system degenerates only after extirpation of the ganglion stellatum, which obviously contains the cells of origin of these perivascular postganglionic fibres. A complete disappearance is, however, only achieved by bilateral removal of the stellate ganglion which shows that the peripheral fibre distribution for the perivascular innervation is crossed and uncrossed and not strictly homolateral.

All this clearly indicates that there are two different types of adrenergic innervation of the inner ear, one strictly perivascular originating in stellate ganglion, and the other independent from blood vessels originating in the superior cervical ganglion (fig. 7).

The functional significance of both systems is not yet clarified. The terminal plexus of the blood vessel-independent adrenergic innervation is in the area of the habenula, where also the initial segments of the cochlear neurons are situated, which suggests a possibly direct adrenergic influence on the initiation of action potentials in the cochlear neurons by changing the threshold of action potential generation. However, such a hypothesis could so far not be confirmed in functional investigation. In a study with behavioural audiometry with cats there were no significant changes in response before and after cervical sympathectomy. An acute stimulation of the cervical sympathicus, however, resulted in an increase of the N_1-amplitude [*Pickles*, 1979]. *Handrock* [1980] on the other hand found a deterioration of the recovery from TTS after sympathectomy. Reports concerning a possible influence of the sympathetic system on the cochlear microphonics are divergent. A sympathetic stimulation leads, according to *Seymour and Tappin* [1953], to an increase but, according to *Bejckert* et al. [1956], to a decrease of the cochlear microphonics, whereas *Baust* et al. [1964] found no effect at all.

The influence of the perivascular adrenergic innervation on the smooth muscle of the blood vessels appears to be rather slow. Stimulation of the stellate ganglion has no short-term effect on the basilar arterial system. A spasm of these arteries, which occurs rather frequently spontaneously or after mechanical stimulation, is not influenced by interruption of sympathetic nerve supply.

There is only little evidence for the existence of a real parasympathetic innervation of the inner ear. In the secretory epithelium of the cristae, unmyelinated nerve fibres are found with enlargements containing great numbers of synaptic vesicles. They could be of parasympathetic nature (fig. 8). To get more information in this question we studied the pathway and origin of all unmyelinated nerve fibres of the VIII nerve, which includes eventual parasympathetic fibres.

In the cochlear nerve there are only extremely few unmyelinated nerve fibres, whereas they are very numerous in the vestibular nerve and its branches. Within the modiolus almost all unmyelinated nerve fibres run together with the intraganglionic spiral bundle, which consists of one third myelinated and unmyelinated fibres. In degeneration studies after various selective nerve lesions the following results were found in the IGSB:

A 50 % reduction of the unmyelinated fibres occurs after extirpation of the superior cervical ganglion, whereas the myelinated fibres remain in normal numbers. A similar reduction of the unmyelinated and an almost

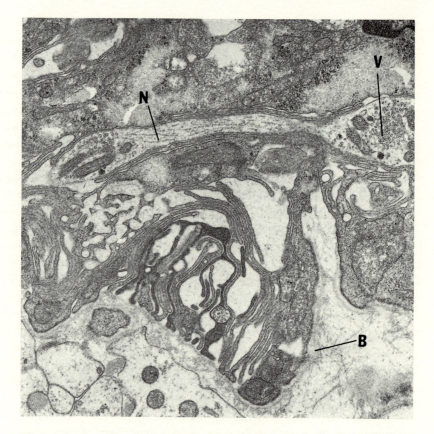

Fig. 8. Basal portion of the dark, so-called secretory epithelium of the crista ampullaris of the cat, showing an intraepithelial unmyelinated nerve fibre (N) with varicosities filled with synaptic vesicles (V). The numerous cytoplasm protrusions towards the basement membrane (B) are typical for this epithelium. Such nerve fibres are indicative of the presence of an autonomic nerve supply of these secretory epithelia.

complete disappearance of the myelinated fibres is found after sectioning of the vestibular nerve. The same results are obtained if the lesion includes the cochlear nerve. The greatest loss with about 90 % disappearance of all myelinated and unmyelinated fibres occurs after a combined lesion of the superior cervical ganglion and the VIII nerve. This shows that about 50 % of the unmyelinated fibres originate in the superior cervical ganglion and the other 50 % in the CNS, reaching the periphery together with the vestibular nerve. Whether these unmyelinated fibres belong to the olivo-cochlear bundle, which according to *Rossi* originates partly in the reticular formation of the brainstem, is finally a question of definition.

References

Arnold, W.: Adrenerge Synapsen im Bereich der Habenula perforata. Archs Oto-Rhino-Lar. *208:* 277–282 (1974).

Baust, W.; Bertucchi, G.; Maruzzi, G.: Changes in the auditory input during arousal in cats with tenotomized middle ear muscles. Archo ital. Biol. *102:* 675 (1964).

Densert, O.: Adrenergic innervation in the rabbit cochlea. Acta oto-lar. *78:* 345–356 (1974).

Falck, B.: Observations on the possibilities of the cellular localization of mono-amines by a fluorescence method. Acta physiol. scand. *56:* suppl. 197 (1962).

Falck, B.; Hillarp, N.-A.; Thieme, G.; Torp, A.: Fluorescence of catecholamines and related compounds condensed with formaldehyde. J. Histochem. Cyto-chem. *10:* 348 (1962).

Handrock, S.: Die Bedeutung des Sympathikotonus auf den temporären und bleiben-den Hörverlust nach Beschallung; Diss. (1980).

Palumbi, G.: Particolare apparato nervoso recettore nella regione apicale della chioc-ciola dell'orecchio humano. Boll. Soc. ital. Biol. sper. *26:* 136 (1950).

Pickles, J. O.: An investigation of sympathetic effects on hearing. Acta oto-lar. *87:* 69–71 (1979).

Seymour, J. C.; Tappin, J. W.: Some aspects of the sympathetic nervous system in-nervation in relation to the inner ear. Acta oto-lar. *43:* 618 (1953).

Spoendlin, H.: Autonomic nerve supply to the inner ear; in Darin de Lorenzo, Vas-cular disorders and hearing defects, pp. 93–111 (University Park Press, Balti-more 1973).

Spoendlin, H.; Lichtensteiger, W.: The adrenergic innervation of the labyrinth. Acta oto-lar. *61:* 423–434 (1966).

Spoendlin, H.; Lichtensteiger, W.: The sympathetic nerve supply to the inner ear. Arch. klin. exp. Ohren Nasen Kehlkopfheilk. *189:* 346 (1967).

Terayama, Y.; Holz, E.; Beck, C.: Adrenergic innervation of the cochlea. Ann. Otol. Rhinol. Lar. *75:* 69–86 (1966a).

Terayama, Y.; Holz, E.; Beck, C.: Adrenergic innervation of the cochlea. Ann. Otol. Rhinol. Lar. *75:* 1–18 (1966b).

Tranzer and Richards, 1971.

Tranzer, J. P.; Thoenen, H.: Electronmicroscopic localization of 5-hydroxydopamine (3,4,5-trihydroxy-phenyl-ethylamine), a new 'false' sympathetic transmitter. Sep-aratum experientia, vol. 23, p. 743 (Birkhäuser, Basel 1967).

Prof. Dr. H. Spoendlin, Vorstand der Universitäts-HNO-Klinik, Innsbruck (Austria)

Adv. Oto-Rhino-Laryng., vol. 27, pp. 14–25 (Karger, Basel 1981)

Autonomic Nervous System and Hearing

B. Maass

Department of Otolaryngology, Justus-Liebig University, Giessen, FRG

According to the current state of morphological knowledge, the inner ear exhibits a remarkably dense adrenergic innervation system. What could be more obvious than to deduce that hearing is, in whatever way, subject to vegetative influences and that dysfunction of the autonomic nervous system could result in disturbances of inner ear function. Thus, all sorts of factors which could possibly affect hearing through an imbalance of the autonomic nervous system have been discussed all along in the etiological consideration of hearing loss. However, in spite of an impressive number of clinical and experimental animal studies, the questions if and how vegetative stimuli could have an effect on the auditory organ are largely unresolved. As is well known, the individual organs are characterized by numerous morphological and functional peculiarities that correspond to their very different specific functions and to their connections with extensive regulation mechanisms. Experimental otology has shown that this also appears to be valid for the auditory organ, which consequently is clearly distinguishable from the central nervous system in the degree of its susceptibility to vegetative influence. In any case, this appears to apply to sympathetic stimulation, inasmuch as it affects cochlear function through its influence on inner ear blood flow. This shall be subsequently discussed in more detail.

For the sake of brevity, reference is made to the literature dealing with the numerous neurovegetative experiments which have been performed to investigate this topic. Due to numerous conflicting reports in the literature, there is no common concept of the susceptibility of inner ear function to sympathetic influence. There may be different reasons for this. The adrenergic innervation of the cochlea exhibits species's, as well as individual variability. The results achieved by working with one species cannot be transferred to another species without consideration of anatomical and functional peculiarities. How one chooses to stimulate or block the sympathetic nervous system is closely related to the choice of the experimental animal. Because of the partial crossing over and interweaving of the fibers, the nervous supply pattern is apparently very complicated. This certainly limits the possibility to interpret experimental results by left-right comparison. The problem is further complicated in that some authors have also described parasympathetic fibers which to some extent are supposed to share a common pathway to the inner ear with adrenergic fibers. Among other considerations, the type of electrical energy and the frequency of stimulation used are important in interpreting stimulation experiments. Inappropriate electrical stimulation can result in a blocking effect due to neuronal damage [40]. Nervous reactions can be substantially influenced by the nature of the anesthesia. This also applies to the susceptibility of peripheral and central circulatory reflexes. Urethane anesthesia has been suggested by some authors to preserve ganglionic transfer within the sympathetic nervous system [27, 36]. On the other hand, experiments involving stimulation of the sympathetic nervous system have demonstrated no difference in results between animals anesthetized with Valium®-Nembutal® and those not undergoing narcosis [45]. Recent investigations also suggest that a sympathetic influence on auditory function can only be demonstrated in awake animals [41]. The dependence of the stimulation response on the autonomic nervous tone is important, and is particularly pronounced in man. It is determined by the sleep-wake rhythm which, in turn, is dependent on superimposed central nervous system functions.

There are suggestions in the literature that an effect after sympathetic nervous system intervention is clearly demonstrated only in a stressed organ. A model based on exposure to noise is appropriate for this purpose. Until now, only a few authors have subjected the inner ear to noise damage prior to investigation of sympathetic influences [11, 14, 25, 26]. According to recent information, the height of the systemic blood pressure is especially important for the demonstration and effectiveness of sympathetic stimuli on inner ear function [18, 30–32]. The general disregard of this very important vegetative parameter, 'blood pressure', is a major disadvantage of previous investigations of sympathetic influence on the inner ear.

Essentially three mechanisms are currently discussed when considering the question how autonomic influences may affect hearing:

(1) The possibility of a direct adrenergic influence on the cell membrane and sensory cell metabolism is taken into consideration by many
authors [5, 6, 12, 46]. A direct sympathetic effect on sensory cell metabolism is currently viewed in connection with the close relationship between noradrenaline and cyclic AMP [9, 58], reputed to play an important role in the central nervous system [17, 47]. Possible connections
between sensory cell metabolism and blood flow are also suggested [5,
9, 58].

(2) According to some authors [10, 55], the strikingly profuse accumulation of adrenergic nerve fiber endings around the non-myelinated
segment of the afferent fibers in the habenular region provides for the
influence of the sympathetic system on the excitation transfer in the afferent neurons and on sound perception through control of the sensitivity
of the auditory organ. The experimental animal studies of *Handrock and
Fischer* [14] also suggest an influence on adaptation behavior. To what
extent the release of transmitter substance into the synaptic space can be
influenced or controlled by the sympathetic nervous system remains, however, to be seen.

(3) The remarkably abundant adrenergic supply to the cochlear vessels in the region of the modiolus, and also, insofar as the inner spiral
vessel is concerned, to segments of the terminal vasculature, suggests that
inner ear function may be sympathetically influenced by way of changes
in blood flow. This mechanism is more thoroughly discussed below.

In contrast to other organs innervated by the autonomic nervous system, most blood vessels are thought to lack an anatomical double innervation [22]. They are rather said to be exclusively innervated by predominantly vasoconstrictive sympathetic nerve fibers. The modulation of
peripheral blood vessel diameter and permeability is mediated through different adrenergic receptors in the area of the arterioles and through local
humoral control mechanisms. It is not yet clear which receptors (α, β) on
the cochlear vessels are of importance [9].

Beyond the muscle containing blood vessels, where adrenergic nerve
fibers, as a rule, are no longer to be found, capillary action mode is regu-
lated by vasoactive metabolites accumulating in the tissues, and is, therefore, dependent upon metabolism [21]. The most important morphological
factors are: the pericytes and endothelial cells, the capillary diameter and
degree of vasodilatation in different capillary regions, the relationship of

one vascular bed volume to another [3, 48–51], and finally, the form or angle of departure of the vessel.

The most distal blood vessels which, from a structural standpoint, are suitable for the regulation of resistance and control of cochlear circulation are the coiled vessels in the wall of the modiolus and the proximal segments of the arterioles which arise from them and are found in the osseous spiral lamina and in the roof of the scala vestibuli [3, 23, 38]. Even if these blood vessels are regarded as being poorly supplied with smooth muscle [38], they nonetheless exhibit a strikingly profuse sympathetic innervation. Like most other vessels, they possess their own tonus, designated as 'tone', which is maintained not only by neuronal pathways, but also through the autonomy of the vessel wall muscle fibers. The latter can be influenced by pressure within the vessel, as well as by the chemical environment. It is not yet clear how strong this tone is in the cochlear vessels. It is important for circulatory regulation and for the distribution of blood within an organ that the vascular threshold of sensitivity to sympathetic stimulation or to circulating catecholamines is dependent on the actual metabolic state. Because acceleration of an organ's metabolic rate raises the threshold for a vasoconstrictive effect, it may be assumed that progressive hypoxia within the cochlea can lead to increasing vascular dilatation. Elimination of the sympathetic innervation further lowers the vascular tone. In our experience, it cannot be assumed that the decrease in tone seen after sympathectomy improves the function of the unstressed cochlea over initial levels. However, one may assume that local chemical control mechanisms coupled with the cellular metabolism considerably contribute to the regulation of intracochlear oxygen partial pressure.

According to recent investigations, the inner spiral vessel assumes a special position among the cochlear vessels with respect to its sympathetic supply [9, 10]. While according to the literature all other capillary regions are free of sympathetic nerve fibers, this capillary vessel still has adrenergic nerve fibers in its surroundings. This raises the question as to how this vessel, reputed to play a leading role in the supply of oxygen to the organ of Corti [1, 24, 28, 29], can be otherwise sympathetically influenced over the arterioles of the modiolus. In this context, several possibilities exist, including influence on vessel permeability [9], control of vessel diameter by endothelial cells and pericytes [15, 16], and local change of blood viscosity [43]. These represent, as yet, little-studied functions and lie outside the scope of this review.

Hemorrhagic hypotension serves as a very useful model for the investigation of vascular autoregulatory capacity. This model should give information about the capacity of the studied vessels to maintain organic blood flow at a constant level in spite of protracted hypotension, and about the extent to which sympathetic adrenergic stimulation after hemorrhage can reduce peripheral circulation through vasoconstriction and sludge formation. The hallmark of hemorrhage is disturbance of the microcirculation. Resulting from a lowering of perfusion pressure, this disturbance precipitates a central nervous system-mediated, sympathetic adrenergic reaction leading to extreme enhancement of sympathetic tone with increased release of catecholamines. A precapillary vasoconstriction is seen, in addition to an increase in blood viscosity in the postcapillary venules. This results in a decrease of capillary perfusion which is synonymous with a decrease of oxygen transport [20]. In our experiments (fig. 1), the fall of intracochlear pO_2 in animals which have not undergone sympathectomy is proportional to the increasing hemorrhagic hypotension. These findings can be interpreted to reflect such a mechanism. Figure 1 clearly shows that measurements on non-sympathectomized animals will miss all behavior regulating pO_2 of a myogenic [4] or mechano-reactive nature [13]. Our oxygen tension measurements have shown that sympathetic innervation has an influence on the mode of reaction of inner ear circulation. After denervation, the fall in O_2 tension that occurs with falling blood pressure during shock is greatly attenuated down to a blood pressure of 65 mm Hg. In this case, an autoregulatory mechanism comes into play. On the other hand, if the innervation of the sympathetic system is intact, the sympathetic system changes the reaction mode of the inner ear circulation so greatly that no autoregulation can be detected up to a blood pressure of 130 mm Hg. One must conclude from this that the inner ear circulation is more dependent on sympathetic innervation and blood pressure than previously assumed. Based on the results of their animal experiments involving cervical sympathetic nervous system stimulation in the presence of various blood pressures, *Hultkrantz* et al. [18] also favor a blood pressure-dependent sympathetic effect on the inner ear circulation. This could also be of clinical significance in the majority of patients suffering from Menière's disease [34] or sudden hearing loss [7, 37] showing hypotensive disturbances of circulatory regulation and in addition exhibit a history of stress. If it is permissible to apply the results of our animal experiments to man, this would mean that hypotension is an important prerequisite for the effect of extreme sympathetic stimulation on

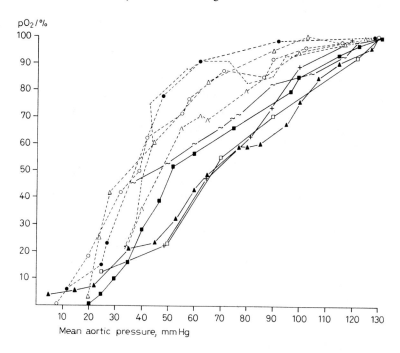

Fig. 1. Behavior of the oxygen partial pressure in endolymph related to mean aortic blood pressure in non-sympathectomized animals (—) and sympathectomized animals (– – –). The pO_2 measurements were made within 20 s after sampling. The pO_2 measured at a mean aortic pressure of 130 mm Hg was considered to be 100 %. Each group included 5 animals [from ref. 32].

diminution of inner ear blood flow, probably through vasoconstriction and sludge formation. Therapeutically, it is therefore necessary to increase perfusion pressure, and with it microcirculation, by optimizing the cardio-vascular condition. Considering the above, this alone would suffice to counteract the increased sympathetic effects on the inner ear vasculature. Alternatively, improvement of the microcirculation by increasing perfusion pressure can also be achieved by a limited local dilatation of the cochlear resistance vessels through conduction anesthesia of the sympathetic nerve, without simultaneous elevation of the blood pressure. At the same time, it ought not to be overlooked that the vagus nerve is often transiently blocked during the occasionally performed sympathetic blockade. This has been shown in animal experiments to have a stabilizing effect on blood pressure through mediation of the cardiac branches of the vagus nerve.

The effect of vagal nerve block during the sympathetic blockade procedure, doubtless beneficial for inner ear blood flow, probably contributes
significantly to the therapeutic usefulness of this procedure and ought not
to be attributed to sympathectomy. Because, under experimental conditions, a sympathetic blockade is only effective on inner ear circulation in
the presence of low blood pressure values, this therapeutic intervention
should be restricted to patients with hearing loss who simultaneously suffer from hypotensive disturbances of circulatory regulation.

Although the effect of sympathectomy on inner ear circulation has
been controversial all along, it has never been doubted that improved circulation occurs in those areas supplied by the carotid artery, such as auri-
cle, tympanic membrane, or mucous membrane of the middle ear. In this
regard, the question arises if sympathetic influences on the cochlea could
not be mediated through capillaries of the tympanic mucous membrane.
In the region of the round window membrane, these capillaries contribute
to the oxygen saturation of the perilymph [33]. However, this question
cannot, as yet, be answered.

The literature is controversial when it comes to effects of noise on inner ear circulation. While some have supported the concept of noise stress
affecting inner ear blood flow [15, 42], others could not confirm such an
influence [2]. *Spoendlin* [56] regards the general state of autonomic reaction as a significant factor in the etiology of damage to human hearing.
Some preliminary investigations of the effects of noise on inner ear blood
flow (fig. 2) suggest that sympathetic blockade decreases the noise-related,
blood pressure-dependent fall in cochlear oxygen tension. The findings also
point to a close relationship between sympathetic effect and blood pressure. In addition to currently unknown factors, the mean aortic blood pressure may significantly affect the effectiveness of the autonomic, especially
sympathetic, influences on inner ear circulation. However, other factors
also appear to determine the extent to which sympathetic stimulation can
affect inner ear blood flow.

An influence of the cervical sympathetic nervous system on the vestibulo-cochlear nucleus and higher centers is often discussed in experimental circles. In this regard, the medial nucleus and vestibulo-cochlear
nucleus are thought to assume particular importance [35, 52, 53]. The
anatomical prerequisite for a sympathetic influence on the cochlear nucleus or reticular formation is the presence of sympathetic plexuses on
the vertebral, i. e. basilar, and posterior inferior cerebellar arteries. These
reach the nuclear area of the eighth cranial nerve over the median branches

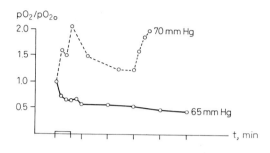

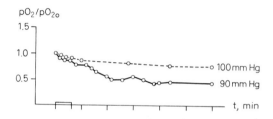

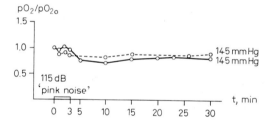

Fig. 2. Behavior of the oxygen partial pressure in endolymph during and after exposure to noise in the (a) hypotensive; (b) normotensive, and (c) hypertensive animal. Sympathectomized side (– – –): control side (—). The pO_2 is related to the basal level (pO_{2o}). The given blood pressures represent the mean of all measurements during the study period. Noise exposure: 115 dB, 'pink noise', for 3 min [from ref. 31].

of the anterior spinal artery [39]. Such a sympathetic influence on the vestibulo-cochlear nucleus and brain stem has been repeatedly proposed by clinicians, partly in connection with the cervical syndrome.

Whether or not cholinergic fibers with actual parasympathetic function exist in the labyrinth in addition to the adrenergic sympathetic fibers is not yet decided, in spite of confirming reports from *Churchill* et al. [8],

Schuknecht et al. [44], and *Suga and Snow* [57]. With respect to their distribution, these fibers, also described by *Ishi* [19], are thought to follow the pathways of the adrenergic fibers [23] as described by *Spoendlin and Lichtensteiger* [54, 55].

References

1   Alford, B. R.; Shaver, E. F.; Rosenberg, J. J.; Guillford, F. R.: Physiologic and histopathologic effects of microembolism of the internal auditory artery. Ann. Otol. *74:* 728 (1965).

2   Angelborg, C.; Hultcrantz, E.; Beausang-Lindner, M.: The cochlear blood flow in relation to noise and cervical sympathectomy. Adv. Oto-Rhino-Laryng., vol. 25, p. 41 (Karger, Basel 1979).

3   Axelsson, A.: The vascular anatomy of the cochlea in the guinea pig and in man. Acta oto-lar. suppl. 243 (1968).

4   Bayliss, W. M.: On the local reactions of the arterial wall to changes of internal pressure. J. Physiol., Lond. *28:* 220 (1902).

5   Beickert, P.; Beck, C.: Der Einfluss vegetativ wirksamer Pharmaka auf das Corti'sche Organ beim Meerschweinchen. Arch. Ohr.-Nas.-KehlkHeilk. *170:* 122 (1956).

6   Beickert, P.: Diskussion zur Ätiologie und Pathogenese des M. Menière. Z. Lar. Rhinol. *41:* 828 (1962).

7   Beickert, P.: Akuter Hörverlust als Notfallsituation. Herz-Kreislauf *6:* 363 (1974).

8   Churchill, J. A.; Schuknecht, H. F.; Doran, R.: Acetylcholinesterase activity in the cochlea. Laryngoscope, St Louis *66:* (1956).

9   Densert, O.: Adrenergic innervation in the rabbit cochlea. Acta oto-lar. *78:* 345 (1974).

10 Densert, O.; Flock, A.: An electron-microscopic study of adrenergic innervation in the cochlea. Acta oto-lar. *77:* 185 (1974).

11 Dietzel, K.; Kleinfeld, D.; Dahl, D.: Klinische und tierexperimentelle Untersuchungen über die Leistungssteigerung des Innenohres nach Stellatumblockade. Wiss. Z. Univ. Halle XX 1971 M, *1:* 24 (1971).

12 Dohlmann, G.: Vegetative Funktionen des inneren Ohres (Labyrinth). Pathophysiologie des vegetativen Nervensystems (Hippokrates-Verlag, Stuttgart 1963).

13 Golenhofen, K.: Die Wirkung von Adrenalin auf die menschlichen Muskelgefässe. 25. Tagung d. dt. Ges. f. Kreislaufforsch., Bad Nauheim 1959, p. 96.

14 Handrock, M.; Fischer, R.: Die Bedeutung des Sympathikotonus auf den temporären und bleibenden Hörverlust nach Beschallung. Z. Hörgeräte-Akustik *18:* 45–52 (1979).

15 Hawkins, J. E., Jr.: The role of vasoconstriction in noise-induced hearing loss. Ann. Otol. *80:* 903 (1971).

16 Hawkins, J. E., Jr.: Microcirculation in the labyrinth. Arch. Oto-Rhino-Lar. *212:* 241 (1976).

17   Hoffer, B. J.; Siggins, G. R.; Bloom, F. E.: Possible cyclic AMP-mediated adrenergic synapses to rat cerebellar Purkinje cells. Adv. Biochem. Psychopharmacol. *3:* 349 (1970).
18   Hultcrantz, E.; Linder, J.; Angelborg, C.: Sympathetic effects on cochlear blood flow at different blood pressure levels. Inserm *68:* 271 (1977).
19   Ishi, T.: Acetylcholinesterase activity in the perivascular nerve plexus of the basilar and labyrinthine arteries. Acta oto-lar. *72:* 281–287 (1971).
20   Jesch, F.; Messmer, K.: Der haemorrhagische Schock – Theoretische Grundlagen. Notfallmedizin *2:* 302 (1976).
21   Johnsson, L. G.: Cochlear blood vessel pattern in the human fetus and postnatal vascular evolution. Ann. Otol. *81:* 22 (1972).
22   Keidel, W. D.: Kurzgefasstes Lehrbuch der Physiologie. 3. Aufl. (Thieme, Stuttgart 1973).
23   Kimura, R. S.; Ota, C. Y.: Ultrastructure of the cochlear blood vessels. Acta oto-lar. *77:* 231 (1974).
24   Kimura, R.: Cochlear vascular lesions; in De Lorenzo, Vascular disorders and hearing defects (University Park Press, Baltimore 1972).
25   Kleinfeld, D.; Dahl, D.: Tierexperimentelle Untersuchungen über das Verhalten der Mikrophonpotentiale nach Stellatumblockade. I. Mikrophonpotentiale nach Stellatumblockade am gesunden Ohr. Arch. klin. exp. Ohr.-Nas.-Kehlk-Heilk. *190:* 124 (1968).
26   Kleinfeld, D.; Dahl, D.: Tierexperimentelle Untersuchungen über das Verhalten der Mikrophonpotentiale nach Stellatumblockade. II. Mikrophonpotentiale nach Stellatumblockade am durch Schall vorgeschädigten Ohr. Arch. klin. exp. Ohr.-Nas.-KehlkHeilk. *190:* 398 (1968).
27   Larrabee, M. G.; Posternak, J. M.: Selective action of anesthetics on synapses and axons in mammalian sympathetic ganglia. J. Neurophysiol. *15:* 91 (1952); cited in Beickert, P.; Gisselsson, L.; Löfström, B.: Der Einfluss des sympathischen Nervensystems auf das Innenohr. Arch. Ohr.-Nas.-KehlkHeilk. *168:* 495 (1956).
28   Lawrence, M.: Effects of interference with terminal blood supply on organ of Corti. Laryngoscope, St Louis *76:* 1318 (1966).
29   Lawrence, M.; Nuttal, A. L.; Burgio, A. B.: Cochlear potentials and oxygen associated with hypoxia. Ann. Otol. *84:* 499 (1975).
30   Maass, B.; Baumgärtl, H.; Lübbers, D. W.: Wirkung einer oberen zervikalen Sympathektomie auf den cochleären Sauerstoffpartialdruck (pO$_2$) unter den Bedingungen einer hämorrhagischen Hypotension. Arch. Oto-Rhino-Lar. *216:* 519 (1977).
31   Maass, B.; Baumgärtl, H.; Lübbers, D. W.: Lokale pO$_2$- und pH$_2$-Messungen mit Mikrokoaxialnadelelektroden an der Basalwindung der Katzencochlea nach akuter oberer zervikaler Sympathektomie. Arch. Oto-Rhino-Lar. *221:* 269–284 (1978).
32   Maass, B.; Baumgärtl, D. W.; Lübbers, D. W.: Wirkung einer Sympathektomie auf den Sauerstoffpartialdruck (pO$_2$) in der Cochlea unter hämorrhagischer Hypotension. Lar. Rhinol. *58:* 665–670 (1979).
33   Maass, B.; Esser, G.: Einfluss des Sauerstoffdruckes im Mittelohr auf die Innenohrfunktion. Lar. Rhinol. *59:* 155–158 (1980).

34   Meyer zum Gottesberge, A.; Stupp, H.: 'Menièresche Krankheit'; in Berendes, Link, Zöllner, Hals-Nasen-Ohrenheilkunde in Praxis und Klinik, vol. 6, chap. 38.7 (Thieme, Stuttgart 1980).

35   Montadon, A.: Fonctions vestibulaires et système neurovégétative; in Monnier, Physiologie und Pathophysiologie des vegetativen Nervensystems (Thieme, Stuttgart 1963).

36   Normann, N.; Löfström, B.: Interaction of D-tubocurarine, ether, cyclopropane and thiopental on ganglionic transmission. J. Pharmac. exp. Ther. *114:* 231 (1955); cited in Beickert, P.; Gisselsson, L.; Löfström, B.: Der Einfluss des sympathischen Nervensystems auf das Innenohr. Arch. Ohr.-Nas.-KehlkHeilk. *168:* 495 (1956).

37   Plath, P.: Schwerhörigkeit bei Herz- und Kreislauferkrankungen. Z. Lar. Rhinol. *56:* 334 (1977).

38   Ritter, K.: Angioarchitektonik und Vasomotion der Gefässstrombahn der Cochlea. Experimentelle Untersuchungen zur Topographie, Morphologie und Funktion der Vascularisation der Meerschweinchenschnecke; Habil.-Schrift, Mainz (1974).

39   Rohr, H.; Unterharnscheidt, F.; Decher, H.: Cochleovestibuläre Störungen bei Halssympathicusschädigungen. Fortschr. Neurol. Psychiat. *28:* 285 (1960).

40   Roll, D.: Ist eine Blockade des Ganglion stellatum durch niederfrequente Reizströme möglich? Z. Allgemeinmed. (Der Landarzt) *45:* 1496 (1969).

41   Rubinstein, M.; Muchnik, C.; Hildesheimer, M.: Effect of emotional stress on hearing. Arch. Oto-Rhino-Lar. *228:* 295 (1980).

42   Schnieder, E. A.: Die Entstehung des Schalltraumas. Ein Beitrag über die Physiologie der Perilymphe. Habil.-Schrift, Würzburg (1970).

43   Schnieder, E. A.: Innenohr- und Hirndurchblutung. Z. Lar. Rhinol. *52:* 186 (1973)

44   Schuknecht, H.; Churchill, J.; Doran, R.: The localization of acetylcholinesterase in the cochlea. Arch. Otolar. *69:* 549 (1959).

45   Sercombe, R.; Aubineau, P.; Edvisson, L.; Grivas, C.; Mamo, H.; Owman, C.; Pinard, E.; Seylaz, J.: CBF changes by sympathetic nerve stimulation and denervation: correlation with the degree of sympathetic innervation of cerebral arterioles. Blood Flow and Metabolism in the Brain, Proc. 7th Int. Symp. on Cerebral Blood Flow and Metabolism, Aviemore 1975.

46   Seymour, J. C.: Observations on the circulation in the cochlea. J. Laryng. *68:* 689 (1954).

47   Singer, G. G.; Goldberg, A. L.: Cyclic AMP and transmission at the neuromuscular junction. Adv. Biochem. Psychopharmacol. *3:* 335 (1970).

48   Smith, C. A.: Capillary areas of the cochlea in the guinea pig. Laryngoscope, St Louis *61:* 1073–1095 (1951).

49   Smith, C. A.: The capillaries of the vestibular membranous labyrinth in the guinea pig. Laryngoscope, St Louis *63:* 87 (1953).

50   Smith, C. A.: Capillary areas of the membraneous labyrinth. Ann. Otol. *63:* 435–447 (1954).

51   Smith, C. A.: Vascular patterns of the membranous labyrinth; in Darin de Lorenzo, Vascular disorders and hearing defects (University Park Press, Baltimore 1972).

52 Spiegel, E. A.; Demetriades, T. D.: Beiträge zum Studium des vegetativen Nervensystems. III. Mitteilung. Der Einfluss des Vestibularapparates auf das Gefässsystem. Pflügers Arch. ges. Physiol. *196:* 185 (1922).

53 Spitzer, A.: Anatomie und Physiologie der zentralen Bahnen des Vestibularis. Arb. neurol. Inst. Univ. Wien *25:* 423 (1924).

54 Spoendlin, H.; Lichtensteiger, W.: The adrenergic innervation of the labyrinth. Acta oto-lar. *61:* (1966).

55 Spoendlin, H.; Lichtensteiger, W.: The sympathetic nerve supply to the inner ear. Arch. klin. exp. Ohr.-Nas.-KehlkHeilk. *189:* 346–359 (1967).

56 Spoendlin, H.: Akustisches Trauma; in Berendes, Link, Zöllner, Hals-Nasen-Ohrenheilkunde in Praxis und Klinik, vol. 6. chap. 42.34 (Thieme, Stuttgart 1980).

57 Suga, F.; Snow, J. B.: Adrenergic control of cochlear blood flow. Ann. Otol. Rhinol. Lar. *78:* 358 (1969).

58 Todd, N. W.; Dennard, J. E.; Clairmont, A. A.; Jackson, R. T.: Sympathetic stimulation and otic blood flow. Ann. Otol. *83:* 84 (1974)

B. Maass, MD, Department of Otolaryngology, Justus-Liebig-University,
D-6300 Giessen (FRG)

Adv. Oto-Rhino-Laryng., vol. 27, pp. 26–39 (Karger, Basel 1981)

Cerebral Blood Flow:
Physiology, Pathophysiology and
Pharmacological Effects

W.-D. Heiss

Forschungsstelle für Hirnkreislaufforschung im Max-Planck-Institut
für Hirnforschung, Köln-Merheim, FRG

As long as a detailed investigation of the circulation of the inner ear
has not been performed and physiological, pathophysiological and pharmacological data of flow to the cochlea and the vestibular organ are scarce,
it may be assumed that the same regulatory mechanisms apply to this system as are acting on cerebral circulation. Therefore, a short survey of
physiology and pathophysiology of cerebral blood flow (CBF) and some
pharmacological results are presented in the following.

Normal Values of CBF and Cerebral Metabolism

When compared to other organs perfusion and metabolism of the
brain are high: The brain weighs approximately 2% of the adult body, but
consumes 20% of the total oxygen uptake and receives 16% of cardiac
output [30]. Related to unit of weight CBF averages 50–55 ml/100 g · min,
but regional cerebral blood flow (rCBF) depends on tissue morphology
and location: in animal experiments rCBF was measured with autoradiographic techniques to be between 138 ml/100 g · min (sensory motor cortex) and 23 ml/100 g · min (white matter) [25, 34, 35]. In normal healthy
man perfusion of gray matter averages 80, that of white matter 20 ml/
100 g · min [12, 16, 19, 20, 42, 44]. As demonstrated in figure 1, marked
regional differences exist with a maximal flow in the frontal lobe [21],
high flows in the temporal pole and close to the Sylvian fissure, while the
lowest values are found in parietal, occipital and posterior temporal regions. These regional heterogeneities were related to the higher activity of

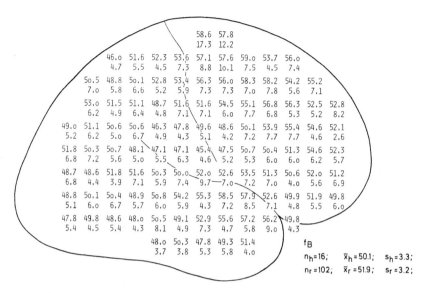

Fig. 1. Mean values and standard deviations of rCBF from 16 patients without lesions of the CNS. Flow is highest over the upper frontal region and the temporal pole, above mean in lower frontal, insular and upper parietal region, below, mean in other parietal and temporal areas and occipital lobe.

especially the frontal lobe during rest and to the higher portion of gray matter in the better-perfused areas.

Under normal conditions the energy requirement of the brain (17 cal/100 g · min) is covered by aerobic glycolysis: 3.7 ml oxygen/100 g · min is needed to metabolize 5.5 mg glucose/100 g · min. 92 % of glucose is metabolized to CO_2 and H_2O, 8 % is converted to lactic acid and pyruvate by anaerobic glycolysis.

CBF, consumption of oxygen and glucose change during life: at the age of 5 years the average CBF doubles that of an adult [23] which is constant up to the age of 50 years. Above 50 years flow usually is slightly reduced even in subjects without central nervous system disease or vascular disorders [29, 32]. Glucose consumption behaves similarly. The relationship between flow and glucose consumption indicates an important principle of CBF regulation: flow is coupled on the metabolic demand of the tissue. Under normal conditions, metabolism and flow of the brain are kept constant and vary insignificantly during physiologic alterations.

CBF Regulation

According to the law of Hagen and Poiseuille the flow (I) in a rigid tube is proportional to the pressure (p), the radius (r) and inverted proportional to the length (l) of the tube and the viscosity (η) of the liquid:

$$I = \frac{p \, r^4 \, \pi}{8 \, l \, \eta} \, .$$

The length of the vessels is an anatomical determinant and will not be considered further on. Of utmost influence on flow are changes in the diameter of the vessels because they are related to flow with their 4th power: Doubling of r increases flow 16-fold, changing r by 20% doubles perfusion. Various regulatory mechanisms are capable of keeping CBF constant during physiologic alterations and of adapting CBF to the momentary tissue demands.

Autoregulation is the ability of cerebral vessels to alter vascular resistance by changing the diameter for keeping flow constant despite varying perfusion pressure. As demonstrated in animal experiments autoregulation keeps flow constant for a mean blood pressure range between 40 and 160 mm Hg [22, 26]. *Strandgaard* et al. [41] have shown that autoregulation is also acting in normal subjects: flow is constant within the autoregulatory range. If mean arterial blood pressure is decreased below a critical threshold (60–80 mm Hg) flow is diminished because a further decrease of vascular resistance cannot be achieved in maximally dilated vessels: below this critical value flow is dependent on perfusion pressure. If blood pressure is raised above the upper limit of autoregulation (150–200 mm Hg) 'breakthrough' of autoregulation occurs: vessels maximally constricted cannot further resist the intravascular pressure; the passive dilatation of arteriols and arteries leads to an overproportional increase of flow. The passive vasodilatation does not occur simultaneously in all parts of the vessels: some segments are dilated while others are still constricted, leading to the 'sausage string' phenomenon [5].

As a clinical consequence the increase of blood pressure above the upper limit of autoregulation may lead to an *acute hypertensive encephalopathy* (crisis) with the symptoms headache, nausea, vomiting, disturbance of consciousness, focal neurological disorders and epileptic seizures. The old concept of such hypertensive encephalopathy took constrictions of cerebral arteries for the pathogenetic mechanism. The new concept of pathologic vasodilatation considers the overextension of the vessel wall as the important pathogenetic factor causing disruption of the tight junctions between the endothelial cells which leads to disturbance of the blood-brain barrier and exsudation of proteins with brain edema.

Hypertension is dangerous because it leads to acute hypertensive encephalopathy but the increased blood pressure may also cause microaneurysms whose rupture may be followed by a disastrous intracerebral hemorrhage. Additionally, the chronically raised blood pressure causes also a hyperplasia of the vessel wall at the cost of the vessel lumen. The thickened vessel wall cannot react to autoregulatory needs as does the normal vessel: the curve of autoregulation is shifted upwards [41]. Due to the resulting higher pressure values necessary for breakthrough of autoregulation, chronic hypertensive patients are not so endangered by sudden increases in blood pressure. But also the lower limit of autoregulation is shifted upwards if mean arterial blood pressure is decreased to below 100–120 mm Hg, CBF diminished in such patients. Therefore, these patients with hypertrophic and fibrotic vessel walls are more endangered by decreases of blood pressure which may lead to hypoxic tissue damage. This interaction explains the high risk of hypertonic patients for ischemic cerebrovascular disease and stresses the importance of a strict therapy of hypertension. By such treatment the risk of stroke in hypertensives is diminished [10]. The consequent treatment of hypertension performed during the last decades may have led already to a decline in the incidence of stroke reported in epidemiologic studies in the USA [190 strokes per 100,000 population in the years 1945–49 compared to 104 in 1970–74; 6].

CO₂ Reactivity. Energy is supplied to the brain mainly by aerobic glycolysis converting glucose to H_2O and CO_2. The higher the metabolic activity of the brain tissue, the more CO_2 is produced and the more blood flow is necessary to transport glucose to and CO_2 away from the tissue. A change in arterial partial pressure of carbon dioxide (pCO_2) causes a proportional alteration of CBF [11]. The CO_2 dependency of flow is given by an S-shaped curve with a linear relationship between 20 and 70 mm Hg and maximal values reached at 70 mm Hg. Between 20 and 70 mm Hg a change of pCO_2 by 1 mm Hg leads to an approximate flow alteration of 2.5 %. If pCO_2 is above 70 mm Hg vasodilatation is maximal and autoregulation is not functioning. Then flow is only dependent on blood pressure.

It was assumed that autoregulation and CO_2 reactivity are based on the same mechanism, namely the changes of pH in the vascular smooth muscle. Changes of pCO_2 are always combined to changes in pH. Autoregulation could only be indirectly related to pH: a fall in blood pressure would lead to a decrease of flow. This flow decrease would cause an increase in tissue pCO_2 followed by a change in pH, which again would dilate the vessels. Against this hypothesis is the fact that CO_2 reactivity and autoregulation may be disturbed independently.

Impairment of Autoregulation and CO_2 Reactivity. (a) After ischemia and hypoxia CO_2 reactivity and autoregulation are lost, the vessels are maximally dilated, paralyzed, perfusion is dependent on perfusion pressure. Hyperemia is found frequently which is of no nutritional value for the destroyed brain tissue. For that state the term 'luxury perfusion' [27] was coined, because flow is far above the needs of the tissue. (b) In states of increased intracranial pressure, e. g. after traumatic brain injury, autoregulation is impaired: perfusion pressure, i. e. the difference between blood pressure and intracranial pressure, is diminished, and blood flow decreases. In such cases CO_2 increase sometimes leads to a vasodilatation, or an increase of blood pressure may improve blood supply. (c) After ischemia autoregulation may recover at a time when CO_2 reactivity is still impaired (postischemic hypofusion). Flow is low and cannot be influenced by changes of blood pressure or pCO_2.

Functional regulation is the coupling of flow on the metabolic demands according to the activity of the brain. The old concept based on measurement of flow and metabolism of the whole brain that these parameters do not change with brain work was overruled by new results showing the dependency of metabolism and flow on functional activity of the nervous tissue. It was demonstrated by autoradiographic methods that glucose consumption is higher in gray than in white matter and related to function [39]: when one eye is enucleated in the rat, the contralateral superior colliculi, lateral geniculate body and striate cortex have significantly decreased glucose uptake [24]. In the rhesus monkey the columnal organization of the striate cortex becomes visible with desoxyglucose autoradiography after unilateral enucleation. The coupling between function and flow/metabolism can be shown in many experimental setups, e. g. also by recording activity of single cortical neurons, electrocorticogram and focal flow before and during epileptic seizures [17]. This coupling may be also observed in humans by means of rCBF measurements with intracarotid xenon: the perfusion of small brain regions is increased during regional activation by special tasks, e. g. motoric activity is followed by increased flow in precentral gyrus, listening to music by flow increase in the temporal lobe, and strong sensory stimulations lead to flow increases in corresponding areas [28]. This technique may also be used for functional mapping to demonstrate the cortical representation of various tasks. The coupling is also present before and during focal and generalized epileptic seizures. The mechanism of this coupling, however, is not clear: mediators may be acid metabolites (change in pH); change in extracellular ionic concentration, especially increase of potassium-ion concentration which leave the cell body during nervous activity; other metabolites liberated during nervous excitation and dilating cerebral vessels, for instance, adenosine [for reviews see 17, 33].

Dependency of Nervous Function on CBF. When flow decreases below 18 ml/100 g · min the activity of neurons in the feline cortex ceases, and cortical neuronal discharge does not return as long a flow stays at or below that threshold. A recovery of flow within a short, up to now not clearly defined, time period to levels above that threshold is followed by resumption of neuronal activity, sometimes even with a long delay [14]. A short-lasting stop of brain circulation with a rapid recirculation at a neuronal level causes only short-lasting changes in the activity of cortical cells. The found threshold of 18 ml/100 g · min is in accordance with the thresholds reported for cessation of EEG activity in man [36, 43] and for recording of evoked potentials in baboons [3]. If flow is decreased below 6–10 ml/100 g · min potassium leaves the cells, indicating the irreversible damage of membranes and cells [1]. These thresholds have important therapeutic implications: only in tissues whose blood supply is not reduced below those thresholds a recovery of function is feasible.

Drug Effects on CBF

The effect of pharmacological agents on CBF is limited by the regulation of cerebral vessels and by the blood-brain barrier, which most drugs cannot penetrate. Dosage-effect relationships found in vessels of other organs are not valid for cerebral arteries. It is beyond the limits of this communication to cover extensively the pharmacology of cerebral circulation dealt with in several reviews [2, 4, 8, 18, 38]. In the following, our own results about effects of drugs on CBF measured by the Xe-washout method are reported [13, 15]. As shown in table I, only a few drugs are

Table I. Drug effects on hemispheric blood flow

Drug	Dosage	Application	Number of measurements	CBF ml/100 g/min	Mean change %	t	p
Control group			37	37.6 ± 7.3	− 3.9 ± 11.1		
Compounds related to chemical mediators							
Midodrine	0.3 mg/kg	i. v.	17	34.2 ± 7.4	+ 5.4 ± 9.2	2.999	0.01
Nylidrin	0.07 mg/kg	i. v.	16	33.1 ± 6.1	− 7.2 ± 12.0	0.962	–
Isoxsuprine	0.07 mg/kg	i. v.	18	34.1 ± 5.8	− 2.9 ± 17.1	0.277	–
Ephedrine-xanthines	+42.8 mg ephedrine +87.4 mg theophylline 129.4 mg etofylline	i. v.	4	40.1 ± 10.7	−21.1 ± 7.6	2.993	0.01
Vasoactive drugs							
Etofylline	1.5 mg/kg	i. v.	8	37.3 ± 5.0	− 3.7 ± 16.9	0.046	–
Xanthinol-niacinat	5 mg/kg	i. v.	5	33.7 ± 5.2	− 5.8 ± 8.7	0.036	–
Raubasin	0.14 mg/kg	i. v.	13	37.8 ± 10.2	− 4.1 ± 9.2	0.056	–
Drotaverine	0.6 mg/kg	i. v.	12	36.2 ± 9.1	+ 0.3 ± 14.9	0.903	–
Proxazole CVD group	80 mg	i. v.-inf.	25	33.1 ± 6.3	+ 2.0 ± 10.6	2.098	0.05
non CVD group	80 mg	i. v.-inf.	6	40.4 ± 7.2	+ 2.5 ± 8.5	0.299	–
	40 mg	i. v.	4	33.7 ± 2.8	+ 0.5 ± 6.7	0.767	–
Bencyclan	150 mg	i. v.-inf.	8	30.4 ± 4.6	−11.8 ± 8.9	1.890	–
	80 mg	i. v.	8	34.5 ± 4.0	− 4.9 ± 8.3	0.238	–
Naftidrofuryl	400 mg	i. v.-inf.	9	32.5 ± 5.7	− 7.2 ± 7.8	0.835	–
Vincamine	30 mg/40 min	i. v.-inf.	12	37.7 ± 6.9	− 3.2 ± 7.1	0.198	–
	40 mg/40 min	i. v.-inf.	14	36.7 ± 5.8	+ 4.5 ± 5.3	2.680	0.01

Table I (continued)

Drug	Dosage	Application	Number of measurements	CBF ml/100 g/min	Mean change %	t	p
Tinofedrin	0.2 mg/kg	i. v.-inf.	12	36.2 ± 4.5	− 5.2 ± 7.5	0.373	–
Hexobendine	0.15 mg/kg	i. v.	9	35.9 ± 6.6	+ 4.5 ± 6.5	2.172	0.05
	6–9 mg/kg	p. o.	5	37.2 ± 9.8	+ 0.8 ± 13.1	0.898	–
	0.15 mg/kg + 1.5 mg/kg etofylline + 0.7 mg ethamivan	i. v.	35	33.4 ± 7.5	+ 2.7 ± 10.9	2.554	0.02
Gingko biloba	35 mg	i. v.	12	32.4 ± 7.2	+ 8.4 ± 7.4	3.550	0.001
Central stimulants and metabolic activators							
Ethamivan	0.7 mg/kg	i. v.	3	32.9 ± 9.4	− 2.8 ± 5.4	0.168	–
CDP-choline	400 mg	i. v.	6	32.7 ± 4.7	− 1.9 ± 8.2	0.545	–
Hemodiluting and dehydrating agents							
Dextran 40	500 ml	i. v.-inf.	24	32.3 ± 7.3	+ 6.9 ± 8.2	4.098	0.001
Dextran 40-sorbitol	500 ml	i. v.-inf.	12	29.8 ± 7.4	+ 6.6 ± 15.5	2.641	0.02
Carboanhydrase inhibitor	3 mg/kg	i. v.	5	40.8 ± 5.2	+11.0 ± 14.5	2.720	0.01
	50 mg	p. o.	7	29.2 ± 5.6	+ 3.7 ± 5.4	1.748	–
Cardiac glycosides							
Ouabain 15 min	0.25 mg	i. v.	18	35.8 ± 6.2	+ 6.4 ± 6.8	3.630	0.001
Ouabain 90 min	0.25 mg	i. v.	18	35.8 ± 4.1	+ 3.9 ± 6.9	2.710	0.01
Digoxin	0.25 mg	i. v.	18	37.7 ± 6.3	− 3.7 ± 8.8	0.064	–
	0.75 mg	i. v.	18	37.3 ± 6.2	− 5.2 ± 10.5	0.394	–
Methylproscillaridin	1 mg	i. v.	10	42.7 ± 8.8	− 7.5 ± 5.5	0.977	–

able to affect flow to the whole brain. For the evaluation of drug effects on cerebral circulation the spontaneous flow changes occurring between subsequent measurements without drug application have to be taken into consideration. In comparison to such spontaneous alterations in a control group measured twice without drugs a significant increase of hemispheric blood flow was observed after the α-receptor stimulator midodrin, the papaverine-like drug proxazole, the central vasodilator hexobendine alone and in combination, the Vinca alkaloid vincamine in a dosage of 40 mg given within 20 min, an extract from *Gingko biloba,* an inhibitor of carbonic anhydrase, hemodilution with low molecular weight dextran alone and in combination with sorbitol, and the cardiac glycoside ouabain. Ephedrine combined with xanthines caused a significant reduction of CBF.

Effects on rCBF were evaluated by comparing two-dimensional flow maps obtained before and after drug application. For the statistical analysis of drug effects on rCBF a regression analysis of the flow values measured before and after drug application was performed and the regression lines were compared to that obtained in the control group. The reactions of rCBF to drugs were classified according to the patterns observed. Various reaction patterns, diagrammatically presented in figure 2, were seen:

No Effect. rCBF maps after drug application were identical to those before drug application. The regression line found with the drug was not different from the control line. No effect on rCBF was the most frequent response after midodrine, ethamivan, CDP-choline, naftidrofuryl, tinofedrin, digoxin and methylproscillaridin. It was often seen with drotaverine, xanthines, buphenine, isoxsuprine and sometimes with the other drugs.

Homogeneous Reaction. Change of rCBF in the same direction in all regions, expressed as a parallel shift of the regression line.

(a) A homogeneous increase of rCBF within the ischemic focus and in the surrounding was obtained after hemodilution with dextran and dextran-sorbitol, after hexobendine, ouabain, extract of *Gingko biloba,* and carbonic anhydrase inhibitor, and sometimes also after proxazole and vincamine.

(b) A homogeneous decrease of rCBF in the ischemic focus and the surrounding with a shift of the regression line to the bottom was observed in all cases receiving ephedrine-xanthines. Such diffuse flow decreases were also seen after buphenin, isoxsuprine and bencyclan in patients with severe arteriosclerosis.

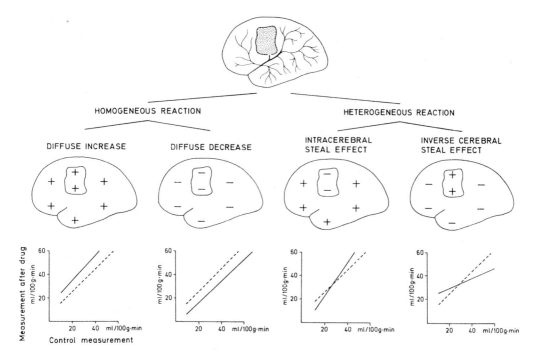

Fig. 2. Drug-induced reaction patterns of rCBF. The various responses are shown as they are found within the ischemic region (encircled) and the surrounding brain. + indicates an increase, − a decrease of rCBF after drug application. At bottom are the graphs obtained by regression analysis: the unbroken line represents the effect of the drug and is compared to that obtained in control measurements without drug application (interrupted line).

Heterogeneous Reaction. Flow in ischemic focus is changed in another direction than flow in surrounding brain. This response is dependent on the resting value of flow and was often not accompanied by alteration of total hemispheric flow.

Intracerebral Steal Phenomenon. Flow to relatively well-perfused regions was improved, while flow to ischemic areas was further impaired. The regression line obtained with this reaction is rotated counterclockwise with respect to control line, the slope of the line is greater than 1. In our studies such a steal reaction was only found after i. v. injection of the central vasodilator hexobendine in patients shortly after their stroke, and only when obstructions of neck vessels were present. Due to the disturbed autoregulation within the ischemic focus during the first days after the attack and the limited total blood supply to the brain by the obstructed vessels the distribution of blood is changed, because resistance in vessels reacting to the vasodilator is decreased and thereby blood is shunted from the ischemic areas where vessels cannot be further dilated.

Inverse Steal Phenomenon. The increase of flow in badly supplied areas is accompanied by a more or less pronounced reduction of flow in better-perfused areas. The regression line is rotated clockwise with respect to the control line, the slope is decreased. Such a response was observed after etofylline, proxazole, vincamine, hexobendine and after low molecular weight dextran. The mechanisms leading to such an inverse steal effect are manifold: Vasoactive substances may open up collateral channels to the ischemic focus and could influence constricted vessels in the neighborhood of the ischemia. With proxazole an additional antiserotonergic action could accentuate this effect. Xanthines have a primary constricting effect on cerebral arteries, which can act on normal vessels but not on paralyzed arteriols in the ischemic focus: The redistribution of blood according to regional changes of vascular resistance brings more blood to the ischemic region. The inverse steal by hemodiluting agents may be related to the decrease in viscosity which is especially effective in low flow areas and when aggregation of blood elements occurs; a dehydration may additionally decrease vascular resistance by an action on perifocal edema.

It has to be stressed, however, that the various reaction patterns observed in single patients and statistically analyzed in groups of patients only show the efficacy of a drug under experimental conditions; their clinical usefulness in the treatment of patients with cerebrovascular diseases has to be proved in controlled clinical studies. For low molecular weight dextran which increases CBF and is used extensively for the treatment of stroke patients only one retrospective study demonstrated improvement [9], while other prospective studies did not yield significant differences between treated and control patients [7, 31, 40]. With other drugs, e. g. vincamine, the results are similar: several investigations indicated its action on CBF, but the convincing clinical evidence is still lacking. This might be due to pharmacokinetic properties of vincamine: an increase of CBF was only observed when dosages were applied resulting in plasma levels above 500 ng/ml [37]; if these levels were not reached, e. g. by the usual oral administration, CBF was not improved. The concept of sufficient plasma levels is fully accepted for other treatments, e. g. antibiotics, cardiac glycosides, and should also be applied to the therapy of cerebrovascular disease.

Summary

Cerebral blood flow (CBF) is high (mean of 50 ml/100 g · min) to cover metabolic and energy requirements of the brain. Independent from blood pressure in the systemic circulation CBF is kept constant by autoregulation, but it reacts to changes of arterial and tissue pCO_2 and to metabolic needs of brain tissue resulting from functional activation. Below defined flow thresholds the function of the nervous tissue is abolished and its morphological integrity destroyed. Due to the regulatory mechanisms only a few drugs are able to affect CBF. The effects depend on the resting blood supply of small regions; this may lead to heterogeneous reaction patterns. The knowledge of drug effects on regional flow may be important in planning a treatment of cerebrovascular disease.

References

1   Astrup, J.; Symon, L.; Branston, N. M.; Lassen, N. A.: Cortical evoked potential and extracellular K^+ and H^+ at critical levels of brain ischemia. Stroke *8:* 51–57 (1977).

2   Betz, E.: Pharmakologie des Gehirnkreislaufs; in Der Hirnkreislauf, pp. 411–440 (Thieme, Stuttgart 1972).

3   Branston, N. M.; Symon, L.; Crockard, H. A.; Pasztor, E.: Relationship between the cortical evoked potential and local cortical blood flow following acute middle cerebral artery occlusion in the baboon. Expl Neurol. *45:* 195–208 (1974).

4   Carpi, A.: Pharmacology of the cerebral circulation (Pergamon Press, Oxford 1972).

5   Farrar, J. K.; Jones, J. V.; Graham, D. I.; Strandgaard, S.; MacKenzie, E. T.: Evidence against cerebral vasospasm during acutely induced hypertension. Brain Res. *104:* 176–180 (1976).

6   Garraway, W. M.; Whisnant, J. P.; Furlan, A. J.; Phillips, L. H., II; Kurland, L. T.; O'Fallon, W. M.: The declining incidence of stroke. New Engl. J. Med. *300:* 449–452 (1979).

7   Gilroy, J.; Barnhart, M. I.; Meyer, J. S.: Treatment of acute stroke with dextran 40. J. Am. med. Ass. *210:* 293–298 (1969).

8   Gottstein, U.: Der Hirnkreislauf unter dem Einfluss vasoaktiver Substanzen (Hüthig, Heidelberg 1962).

9   Gottstein, U.; Sedlmeyer, I.; Heuss, A.: Behandlung der akuten zerebralen Mangeldurchblutung mit niedermolekularem Dextran: Therapie-Ergebnisse der retrospektiven Studie. Dt. med. Wschr. *101:* 223–227 (1976).

10 Gottstein, U.: Zerebrale Haemodynamik bei arteriellem Hochdruck und Hoch-

druckkrise sowie unter dem Einfluss therapeutischer Blutdrucksenkung. Verh. dt. Ges. KreislForsch. *43:* 61–74 (1977).

11   Harper, A. M.; Glass, H. I.: Effect of alterations in the arterial carbon dioxide tension on the blood flow through the cerebral cortex at normal and low arterial blood pressures. J. Neurol. Neurosurg. Psychiat. *28:* 449–452 (1965).

12   Heiss, W.-D.; Prosenz, P.; Roszuczky, A.; Tschabitscher, H.: Die Verwendung von Gamma-Kamera und Vielkanalspeicher zur Messung der gesamten und regionalen Hirndurchblutung. Nucl. Med. *7:* 297–318 (1968).

13   Heiss, W.-D.: Drug effects on regional cerebral blood flow in focal cerebrovascular disease. J. neurol. Sci. *19:* 461–482 (1973).

14   Heiss, W.-D.; Hayakawa, T.; Waltz, A. G.: Cortical neuronal function during ischemia. Effects of occlusion of one middle cerebral artery on single unit activity in cats. Archs Neurol., Chicago *33:* 813–821 (1976).

15   Heiss, W.-D.: Effects of drugs on cerebral blood flow in man. Adv. Neurol. *25:* 95–114 (1979).

16   Heiss, W.-D.: Regional cerebral blood flow measurement using a scintillation camera. Clin. nucl. Med. *4:* 385–396 (1979).

17   Heiss, W.-D.; Turnheim, M.; Vollmer, R.; Rappelsberger, P.: Coupling between neuronal activity and focal blood flow in experimental seizures. Electroenceph. clin. Neurophysiol. *47:* 396–403 (1979).

18   Herrschaft, H.: Gehirndurchblutung und Gehirnstoffwechsel. Fortschr. Neurol. Psychiat. *44:* 195–322 (1976).

19   Hoedt-Rasmussen, K.: Regional cerebral blood flow: the intraarterial injection method. Acta neurol. scand. *43:* suppl. 27, pp. 1–79 (1967).

20   Ingvar, D. H.; Cronqvist, S.; Ekberg, R.; et al.: Normal values of regional cerebral blood flow in man, including flow and weight estimates of grey and white matter. Acta neurol. scand. suppl. 14, p. 72 (1965).

21   Ingvar, D. H.; Philipson, L.; Torlöf, P.; et al.: The average rCBF pattern of resting consciousness studied with a new colour display system. Acta neurol. scand. suppl. 64, p. 252 (1977).

22   Johansson, B.: Regional cerebral blood flow in acute experimental hypertension. Acta neurol. scand. *50:* 366–372 (1974).

23   Kennedy, C.; Sokoloff, L.; Anderson, W.: Cerebral blood flow and metabolism in normal children. Am. J. Dis. Child. *88:* 813 (1954).

24   Kennedy, C.; Rosiers, M. H. des; Jehle, J. W.; Reivich, M.; Sharpe, F.; Sokoloff, L.: Mapping of functional neural pathways by autoradiographic survey of local metabolic rate with (^{14}C)deoxyglucose. Science *187:* 850–853 (1975).

25   Landau, W. M.; Freygang, W. H.; Roland, L. P.; Sokoloff, L.; Kety, S. S.: The local circulation of the living brain: values in the unanesthetized and anesthetized cat. Trans. Am. neurol. Ass. *80:* 125–129 (1955).

26   Lassen, N. A.: Cerebral blood flow and oxygen consumption in man. Physiol. Rev. *39:* 183–238 (1959).

27   Lassen, N. A.: The luxury perfusion syndrome and its possible relation to acute metabolic acidosis localized within the brain. Lancet *2:* 1113 (1966).

28   Lassen, N. A.; Ingvar, D. H.; Skinhoj, E.: Brain function and blood flow. Sci. Am. *239:* 50–59 (1978).

29   Lavy, S.; Melamed, E.; Bentin, S.; Cooper, G.; Rinot, Y.: Bihemispheric de-

creases of regional cerebral blood flow in dementia: correlation with age-matched normal controls. Ann. Neurol. *4:* 445–450 (1978).

30 Lübbers, D. W.: Physiologie der Gehirndurchblutung; in Der Hirnkreislauf, pp. 214–260 (Thieme, Stuttgart 1972).

31 Matthews, W. B.; Oxbury, J. M.; Grainger, K. M. R.; Greenhall, R. C. D.: A blind controlled trial of dextran 40 in the treatment of ischemic stroke. Brain *99:* 193–206 (1976).

32 Naritomi, H.; Meyer, J. S.; Sakai, F.; Yamaguchi, F.; Shaw, T.: Effects of advancing age on regional cerebral blood flow. Archs Neurol., Chicago *36:* 410–416 (1979).

33 Purves, M. J.: Control of cerebral blood vessels: present state of the art. Ann. Neurol. *3:* 377–383 (1978).

34 Reivich, M.; Jehle, J.; Sokoloff, L.; Kety, S. S.: Measurement of regional cerebral blood flow with antipyrine-^{14}C in awake cats. J. appl. Physiol. *27:* 296–300 (1969).

35 Sakurada, O.; Kennedy, C.; Jehle, J.; Brown, J. D.; Carbin, G. L.; Sokoloff, L.: Measurement of local cerebral blood flow with iodo(^{14}C)antipyrine. Am. J. Physiol. *234:* H59–H66 (1978).

36 Sharbrough, F. W.; Messick, J. M.; Sundt, T. M., Jr.: Correlation of continuous electroencephalograms with cerebral blood flow measurements during carotid endarterectomy. Stroke *4:* 674–683 (1973).

37 Siegers, C.-P.; Heiss, W.-D.; Kohlmeyer, K.: Plasma- und Liquorspiegel von Vincamin nach intravenöser Infusion bei Patienten. Arzneimittel-Forsch. *27:* 1274–1277 (1977).

38 Sokoloff, L.: The action of drugs on the cerebral circulation. Pharmacol. Rev. *11:* 1–85 (1959).

39 Sokoloff, L.; Reivich, M.; Kennedy, C.; Rosiers, M. H. des; Patlak, C. S.; Pettigrew, K. D.; Sakurada, O.; Shinohara, M.: The (^{14}C)deoxyglucose method for the measurement of local cerebral glucose utilization: theory, procedure, and normal values in the conscious and anesthetized albino rat. J. Neurochem. *28:* 897–916 (1977).

40 Spudis, E.; Torre, E. de la; Pikula, L.: Management of completed strokes with dextran 40. A community hospital failure. Stroke *4:* 895–897 (1973).

41 Strandgaard, S.; Olesen, J.; Skinhoj, E.; Lassen, N. A.: Autoregulation of brain circulation in severe arterial hypertension. Br. med. J. *159:* 507–510 (1973).

42 Sveinsdottir, E.; Torlöf, P.; Risberg, J.; Ingvar, D. H.; Lassen, N. A.: Monitoring regional cerebral blood flow in normal man with a computer-controlled 32-detector system. Eur. Neurol. *6:* 228–233 (1971/72).

43 Trojaborg, W.; Boysen, G.: Relation between EEG, regional cerebral blood flow and internal carotid artery pressure during carotid endarterectomy. Electroenceph. clin. Neurophysiol. *34:* 61–69 (1973).

44 Wilkinson, I. M. S.; Bull, J. W. D.; DuBoulay, G. H.; Marshall, J.; Ross Russell, R. W.; Symon, L.: Regional blood flow in the normal cerebral hemisphere. J. Neurol. Neurosurg. Psychiat. *32:* 367–378 (1969).

Prof. Dr. W.-D. Heiss, Forschungsstelle für Hirnkreislaufforschung im Max-Planck-Institut für Hirnforschung, Ostmerheimer Str. 200, D-5000 Köln 91 (FRG)

Adv. Oto-Rhino-Laryng., vol. 27, pp. 40–69 (Karger, Basel 1981)

Sudden Hearing Loss: A Clinical Survey

H. Feldmann

ENT Clinic, University of Münster, FRG

Statistics and Its Shortcomings

Since the publication of the first singular observations by *Citelli* [2] in 1926 and *De Kleyn* [8] in 1944 numerous authors have reported on cases of sudden deafness, which now run up to more than 3,000 [1, 3, 5, 6, 9, 15–17, 19, 22–24, to quote only a few]. The reports include statistics of large groups, the series of *Shaia and Sheehy* [20], for instance, covering 1,220 cases. Obviously there is no lack of observations, yet our knowledge of this disease, if it is a disease of its own, is still rather inadequate.

Looking for general statistical features it can be stated that there is no predominance with regard to sex – male and female patients are equally affected. The same holds true for the right and the left ear. The distribution relating to age is demonstrated by figure 1. It shows a flat summit in the age groups from 20 to 60. Sudden hearing loss seems to be extremely rare with children, but perhaps this is only due to the fact that it is not reported of the very young [18], and therefore passes unnoticed. The average hearing loss is illustrated by a flat curve around 50–70 dB (fig. 2).

These crude statistical features tend to obscure rather than reveal the nature of this disease. One of the deficiencies of such statistics is the fact that they are based on the assumption that sudden deafness is a nosologic entity, which will emerge clearest when all individual features have been eliminated. It seems worth taking the opposite course and demonstrating the variety of clinical data in the individual case and presenting a kind of phenomenology of sudden hearing loss. Here, of course, the question of definition and differentiation arises, and this points to a further shortcom-

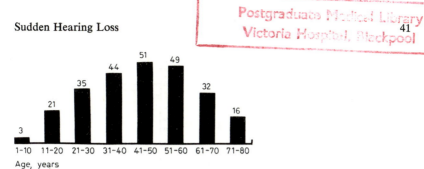

Fig. 1. Age-related distribution of sudden hearing loss, based on 248 observations of *Kemper* [7] and *Wandhöfer* [25].

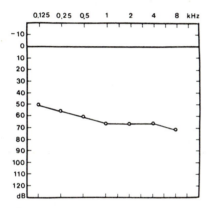

Fig. 2. Averaged hearing loss of 78 cases of *Kemper* [7].

ing of statistics. There is no generally accepted definition, as to what the criteria of sudden deafness are when one speaks of it as a nosological or clinical or etiological entity.

Definition and Its Pitfalls

A definition may serve as a heuristic instrument. Keeping this in mind, one may attempt the following statements: sudden deafness usually is unilateral; occurs in a patient otherwise completely healthy; begins abruptly, suddenly; is sensorineural; has no identifiable cause and is not discernibly connected with other diseases.

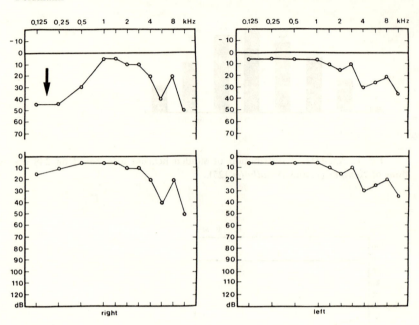

Fig. 3. Sudden hearing loss in a 49-year-old man, prompted by the bang of a toy pistol at a distance of 3 m. Full recovery after 6 days (below). 3 years later a relapse of the sudden hearing loss occurred spontaneously in the same ear without any triggering incidence; again full recovery.

Features which can be determined positively by history and findings are its sudden onset and the character of hearing loss, i. e. sensorineural. Other features must be determined by exclusion: the cause should not be obvious and the hearing loss should not be connected with other diseases. The validity of such statements becomes apparent when they are applied to some real cases.

Causality

With a male patient of 49 the sudden hearing loss (fig. 3) was caused by the bang of a toy pistol fired at a distance of 3 m. Since he had seen that a boy was handling this pistol he was not startled by the shot. Immediately after the bang, however, he noticed that his right ear was deaf.

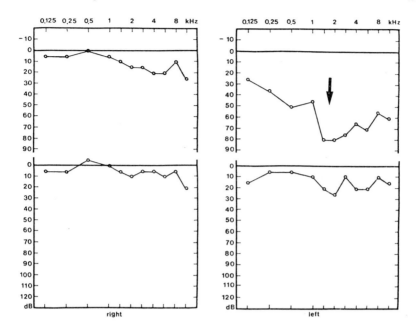

Fig. 4. Sudden hearing loss in a woman of 23, prompted by fright and anxiety. Full recovery within 2 weeks (below). 8 months later there was a relapse of the sudden hearing loss in the same ear, the patient being in the third month of pregnancy, otherwise no unusual circumstances. Again full recovery.

A female of 23 suffered a sudden hearing loss (fig. 4) which was apparently prompted by fright and fear. The patient was startled out of her sleep by a wind gust slamming a door. She was sitting upright in her bed anxious that it might again be a burglar as a few weeks before. In this nervous tension her left ear turned deaf, as though a curtain was drawn. One patient (fig. 5) suffered a hearing loss while diving in a swimming pool to a depth of 3 m; another patient (fig. 6) suffered a sudden deafness while taking a sunbath.

Obviously there are external or internal (nervous) influences, which have prompted sudden hearing loss in individual cases, but they can hardly be considered the exclusive and adequate causes. The last blade of straw breaks the camel's back. This seems to be the appropriate metaphor to describe the effect these circumstances have when bringing about sudden deafness. The ear must have been in a precarious state before, so that the catastrophy could be triggered off by these banal incidents.

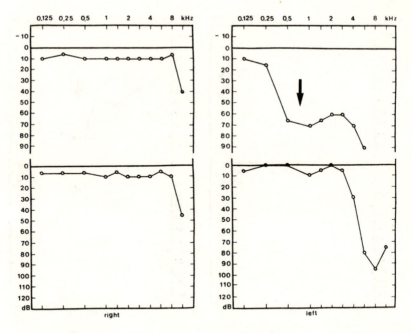

Fig. 5. Sudden hearing loss in a man of 42 when diving in a swimming pool of 3 m depth. No vestibular symptoms, no indication of lesion of round or oval window. Partial recovery (below) under conservative treatment within 9 days.

Mode of Onset

According to the above statements the onset of deafness should be sudden, abrupt. But apparently there are different grades of suddenness. The onset may be a matter of a second. This is often accompanied by a pop-like hearing sensation. Or it may occur within a few minutes, like a curtain being drawn. The hearing loss can fluctuate for hours, or it can develop intermittently taking a few days (fig. 7).

Since this intermittent development is rare it may be illustrated by the following observation (fig. 8). A female of 21 had suffered a sudden hearing loss in the low frequencies in her left ear. Although an intense treatment was started immediately the hearing deteriorated dramatically from day to day, until after 4 days the ear was totally deaf. In this period nystagmus to the left was observed, accompanied by dizziness, but both vestibular organs remained excitable. After 15 days a restitution became noticeable, and after 22 days completely normal hearing had been restored.

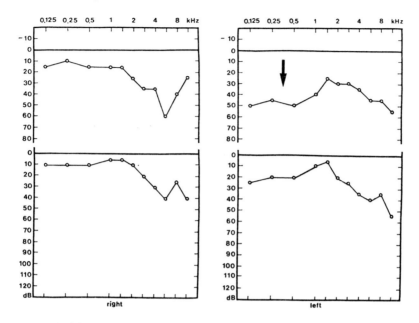

Fig. 6. Sudden hearing loss in a man of 56 during sunbath. Nearly complete recovery within 24 days (below). 4 weeks afterwards relaps of sudden deafness in the same ear; no recovery.

Mode of Progress

After the sudden onset the hearing loss may progress in all conceivable ways (fig. 9): the complete restitution of the normal hearing; the partial restitution; the hearing loss remaining constant, and the progressive deterioration ending in total deafness.

Audiometric Data

According to the above statement the sudden hearing loss is of a sensorineural character. The individual case may present a great variety of audiometric curves. For a systematic approach the following types may be distinguished: low frequency trough (fig. 10); pantonal hearing loss (fig. 11); middle frequency dip (fig. 12); high frequency dip or slope (fig. 13), and total deafness (fig. 14).

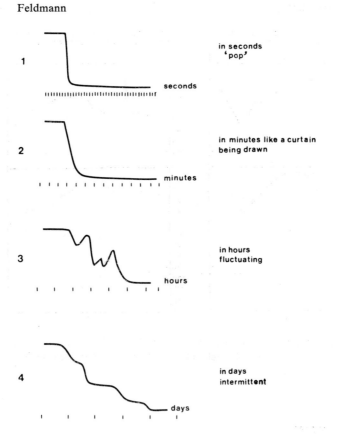

Fig. 7. Different modes of onset of sudden hearing loss.

Fig. 8. Sudden hearing loss in a female patient of 21, audiograms after 1 (above) day and after 2 and 4 days (below) (fig. 8a) show rapid deterioration. Complete deafness from the 5th to the 13th day, associated with slight vestibular disorders. After 14 days beginning recovery in the highest frequencies (fig. 8b above), after 21 days normal hearing fully restored (fig. 8b below).

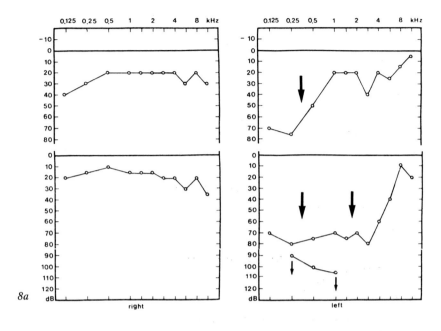

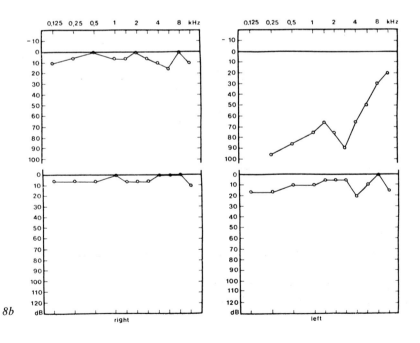

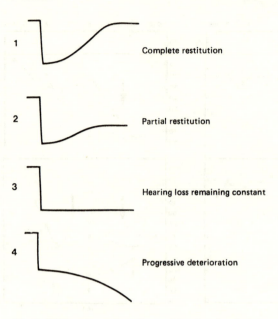

Fig. 9. Different courses after sudden hearing loss.

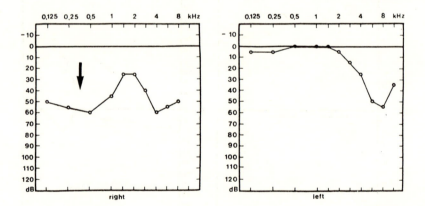

Fig. 10. Low frequency trough. Sudden hearing loss in a male of 34. Complete recovery. 4 months later relapse of same hearing loss, no restitution. Symmetrical noise-induced hearing loss in the high frequencies.

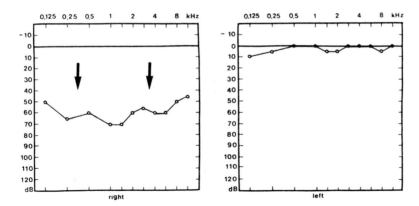

Fig. 11. Pantonal hearing loss. Sudden deafness in a male of 31. No restitution.

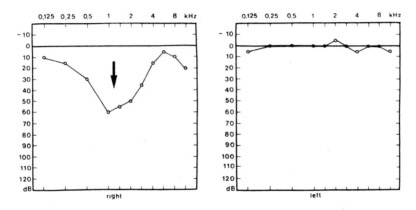

Fig. 12. Middle-frequency dip. Sudden hearing loss in a male of 23. Complete recovery within 6 days.

The data in an audiometric test battery are as manifold as are the threshold curves. There are cases with clear-cut cochlear symptoms and others with symptoms of a retrocochlear lesion. Sometimes the pattern changes during the course.

A male patient of 54 (fig. 15) had suffered a sudden hearing loss in the mid-frequency region. In the acute phase there was no recruitment present, SISI score 0%, Békésy audiogram type IV, extreme threshold decay. There was a complete recovery. During the phase of restitution the

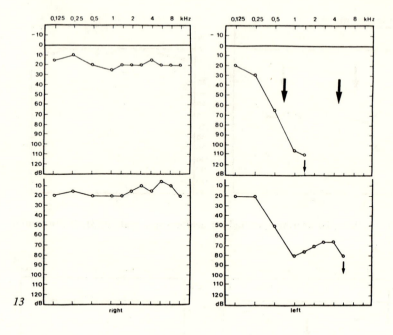

13   right   left

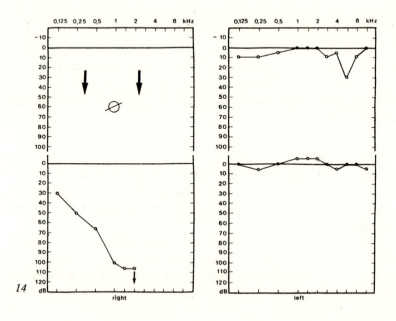

14   right   left

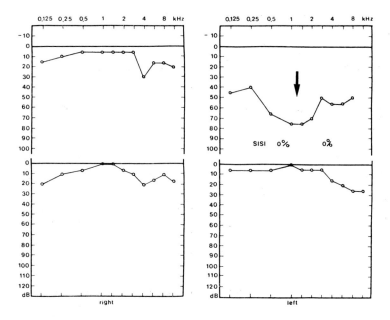

Fig. 15. Sudden hearing loss in a male of 54. SISI score 0 %, pronounced thresh-old decay. Complete recovery within 4 weeks. 1 year later relapse of sudden hearing loss in the same ear, nearly complete deafness, again full recovery.

SISI score became 100 %, recruitment of loudness positive. A testing with regard to tone decay, however, was not repeated.

The observation of high degree tone decay should be a reason to be very careful about applying an audiometric test battery in the acute phase of sudden deafness. In our experience the audiometric test battery is of minor value in the differential diagnosis of sudden deafness; so far the findings can hardly be translated into adequate diagnostic or prognostic terms. The application of higher sound pressure levels, however, is not advisable to an organ which is already on the verge of an ultimate break down.

Fig. 13. High-frequency slope. Sudden hearing loss in a female of 20. Partial recovery within 24 days (below).

Fig. 14. Total deafness. Sudden hearing loss in a girl of 14, slight vestibular symptoms. Partial recovery after 24 days.

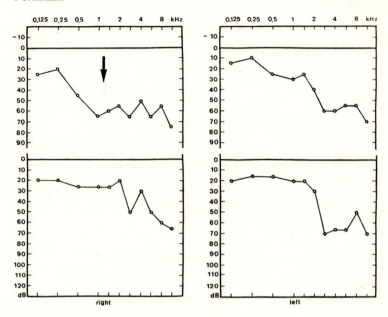

Fig. 16. Sudden hearing loss in a male of 67, presbyacusis and slight noise-induced hearing loss. Restitution to the previous level.

Previous State

If one wants to describe the phenomenology of sudden deafness, it is essential to take into consideration the long-term course and the condition of both ears before the onset of the acute hearing loss. In the cases demonstrated here so far the sudden hearing loss had occurred with patients with two normal ears. The sudden hearing loss, however, can also be an acute deterioration of a previous hearing defect.

A male of 67 with an advanced presbyacusis, perhaps intensified by industrial deafness, suffers a sudden hearing loss in his right ear (fig. 16). There is a complete restitution to the previous state. But there may also have been a defect in the opposite ear prior to the sudden deafness. A 48-year-old male (fig. 17) was suffering from a complete deafness and loss of vestibular function in his right ear, which resulted from a head injury 33 years ago. Now he noticed a sudden deafness in his left ear. Fortunately there was a complete recovery.

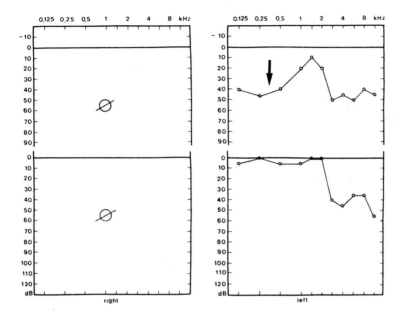

Fig. 17. Male patient of 48, complete deafness and loss of vestibular function in his right ear, following a head injury 33 years ago. Now sudden hearing loss in his last hearing ear. Restitution to the previous level.

The sudden deafness may also involve both ears at the same time. If one systematically relates the possible manifestations of sudden deafness to the state preceding this disease, something like figure 18 results.

Situation 1 is undoubtedly the prototype of sudden deafness: both ears are normal, one ear suffers a sudden loss.

Situation 2, hard of hearing in both ears, sudden deterioration in one ear, was demonstrated in figure 16. It shows the case of a patient suffering from presbyacusis. Nor is this uncommon with children having a hereditary hearing defect, e. g. Pendred's syndrome and other diseases.

Situation 3, sudden deterioration in the poor ear, the opposite ear being normal, is the typical condition with regard to the recurrent sudden hearing loss. As a matter of fact this state probably often passes unnoticed, because the patient will not pay much attention to the deterioration of hearing in his ear which he regards as deaf anyway.

Situation 4, one ear completely deaf, now sudden deafness in the hearing ear, is especially dramatic. The hearing loss in the previously poor

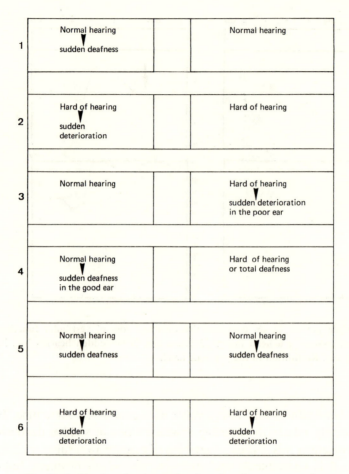

Fig. 18. Sudden hearing loss related to situation at onset of event.

ear may be due to a former sudden deafness or to any other disease. *Lehnhardt* [11, 12] was the first to suggest that this situation might present peculiarities of pathogenesis and prognosis. In analogy to sympathetic ophthalmia the assumption was that an autoaggressive immunologic process is started forming antibodies against the organ of Corti. *Kumpf and Wandhöfer* [10] compared 33 cases of such a consecutive hearing loss in the last hearing ear with 106 'normal' cases of sudden deafness. There was no indication that a sudden deafness in the last hearing ear is a disease of its own, distinguished in any respect from the sudden deafness in the first ear.

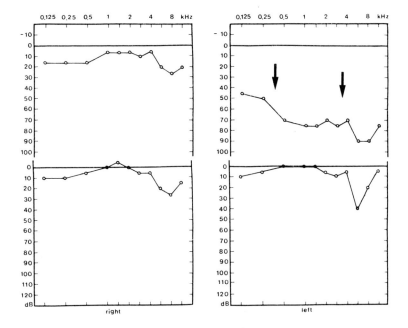

Fig. 19. Male patient of 29. First attack of sudden deafness at the age of 24 in his left ear, second attack at the age of 28 in his right ear, now third attack again in the left ear. Each time full recovery.

Situation 5 and 6, both ears suffering a hearing loss simultaneously, the loss either setting in a symmetric normal or a symmetric hard of hearing condition, must always strongly be suspected to be a symptom of a systemic disease. Examples will be given later on.

Relapse

One important aspect of the long-term course after a sudden deafness is the possibility of a relapse. According to our observations it is rather frequent. Out of the 12 cases demonstrated here so far for various reasons 6 showed one or more relapses in the same ear, partly with restitution, partly ending in permanent hearing loss.

The sudden hearing loss may also recur alternately in the right and left ears. A male patient (fig. 19) suffered a sudden deafness for the first

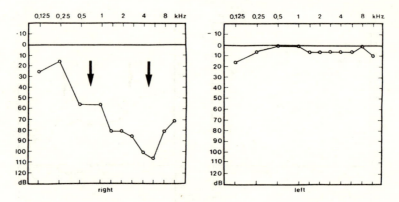

Fig. 20. Female patient of 54. Sudden hearing loss in the right ear, associated with vestibular disorder. No recovery.

time at the age of 24 in his left ear; at the age of 28 there was a sudden deafness in his right ear, and at the age of 29 again a sudden deafness in his left ear. Each time there was a complete recovery.

Concomitant Symptoms

One of the statements about sudden deafness, made at the beginning, was that it should not be connected with other diseases. By including cases that showed hearing problems in one or both ears prior to the sudden deafness, we have, strictly speaking, already broken this rule. It is most likely that a sudden deafness in one ear is related to the state that this ear and the other ear were in before the acute loss. If we take into consideration symptoms different from hearing problems, which had been present prior to the sudden deafness or which have appeared simultaneously with the sudden deafness, the above statement is challenged furthermore.

First and foremost there may be a vestibular disturbance. With a female patient of 54 (fig. 20) the hearing loss in her right ear had occurred suddenly. It was accompanied by vertigo and nausea. The audiometric test battery revealed a retrocochlear lesion. There was no improvement in hearing, the vestibular function remained impaired. An acoustic neuroma is suspected, but could not be confirmed so far. Thus purely symptomatic description 'sudden deafness involving vestibular function' is certainly a most unsatisfying diagnosis.

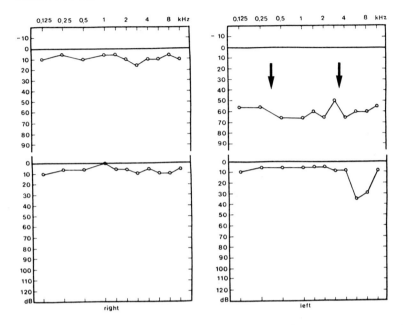

Fig. 21. Male patient of 35. Attacks of sudden hearing loss in his left ear at the age of 29, 31 and 35, each time full recovery. The last event was accompanied by an attack of vertigo, suspect of beginning Menière's disease.

The patient mentioned in figure 21 had suffered recurrent sudden hearing losses in his left ear at the age of 29, 31 and 35, each followed by a full recovery. The last seizure was accompanied by a short attack of vertigo. It is very likely that this will turn out to be a case of Menière's disease. It is not uncommon in Menière's disease to start with a single symptom, e. g. a sudden hearing loss, and it is only after some time that the complete classical syndrome develops.

When both ears are involved simultaneously or consecutively and when there are vestibular symptoms or other concomitant symptoms, a systemic disease may well be supposed.

The patient whose case is demonstrated in figure 22 had a sudden deafness with tinnitus in his left ear, resulting from the bang of a firework at the age of 25. He recovered completely within a few days. 9 days later he suffered a further hearing loss in the same ear, after he had been ex-

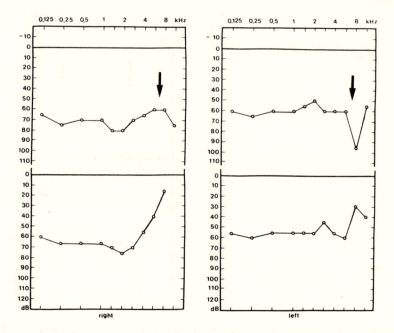

Fig. 22. Male patient of 37. At the age of 25 recurrent attacks of sudden hearing loss under exposition to extreme noise, alternately in both ears. 10 years later sudden deterioration in both ears in the highest frequencies. Final diagnosis: connatal syphilis. Restoration to the previous level by cortisone therapy.

posed to the noise of demolition work. 3 days later he had a similar hearing loss in his second ear while cleaning sheet metal. These occurrences were interpreted as acoustic trauma despite the unusual audiometric findings. At that time he had a bilateral pantonal hearing loss of 60 dB with a steep rise of the curve in the highest frequencies. There were no vestibular symptoms. The hearing could not be restored. All syphilitic reactions available at that time (1969) were negative. 10 years later there was again a sudden hearing loss in the highest frequencies. The patient noticed this at once, because his speech intelligibility depended to a great deal on the high frequencies. This event provided the reason to repeat all the diagnostic procedures. The modern tests, such as FTA-ABS, TPHA and further exploration revealed that it had been a connatal syphilis all the time.

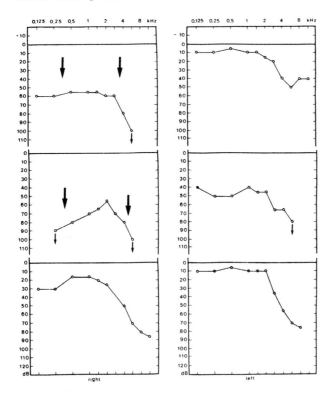

Fig. 23. Female patient of 37. Sudden hearing loss in the right ear, dramatic deterioration, involvement of left ear, complete bilateral loss of vestibular function; audiograms at onset, after 3 months and after treatment with cortisone and 'Imurek'. Final diagnosis: Cogan's disease.

A female patient of 37 (fig. 23) first had a sudden hearing loss in her right ear. In spite of intense treatment the hearing deteriorated dramatically. Then the left ear became involved and within 2 weeks there was a complete loss of vestibular function on both sides. This desperate course made the patient cry frequently. But when her eyes turned red this was not due to crying it was due to a transient keratitis, iritis and conjunctivitis. We diagnosed this as Cogan's syndrome. The difficult diagnosis of Cogan's syndrome may be characterized as 'crying deafness'. By means of a treatment with cortisone and Imurek® the hearing could partly be restored and has been kept stable for more than 2 years now. The vestibular function, however, remains lost in both ears.

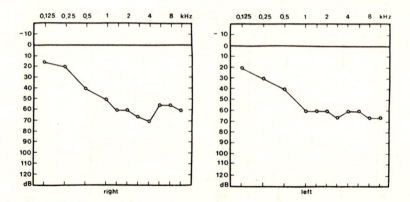

Fig. 24. Female patient of 39. At the age of 36 sudden hearing loss in both ears. No vestibular disorder. Final diagnosis: Refsum's disease, established by high level of phytanic acid in the blood. No improvement of hearing by special treatment.

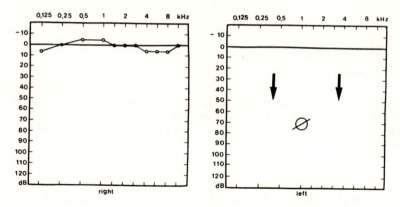

Fig. 25. Sudden deafness in a female of 28, following mumps. No vestibular symptoms. No restitution.

The female patient whose case is demonstrated in figure 24 at the age of 36 suffered a sudden hearing loss in both ears, which started with a click-like sound. Some time before this she had suffered from eye problems and ataxy of gait. Innumerable diagnostic procedures and therapeutic attempts, including a neurosurgical intervention in the occipital cranial fossa, were unsuccessful. We could diagnose this as Refsum's disease and

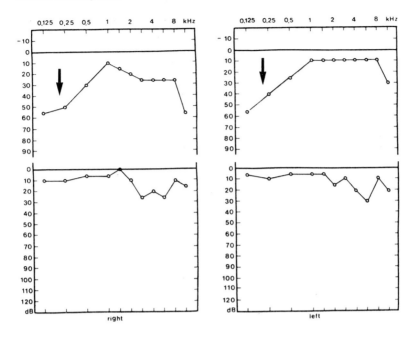

Fig. 26. Male patient of 26, sudden hearing loss in both ears following rubella. Complete restitution.

confirm the diagnosis by the high level of phytanic acid in the blood. Now the patient is being treated with a corresponding diet and plasmaphoresis [4]. In other cases which present a more typical situation with regard to unilateral sudden hearing loss the concomitant circumstances permit an etiological diagnosis.

A female patient of 28 (fig. 25) waking up in the morning noticed that she was completely deaf in her left ear. She suffered from tinnitus and slight dizziness. The vestibular function, however, was normal. She was having the mumps. In the beginning the right parotid gland had been swollen, 1 week later the left parotis became involved. In this phase 1 week after the onset of the first symptoms, the deafness occurred. The diagnosis was confirmed by serologic findings. Unfortunately the hearing could not be restored.

A male patient of 26 (fig. 26) fell ill with rubella. A related pediatrician established the diagnosis, which was unusual for this age. Later on his diagnosis could be confirmed by the high level of antibodies and

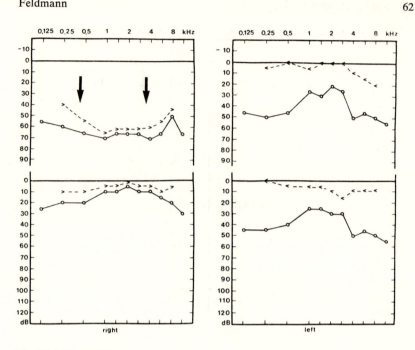

Fig. 27. Male patient of 37 suffering from Waldenstrom's disease (macroglobu-linemia). Sudden hearing loss in his right ear. Complete recovery (below). Chronic otitis media in his left ear.

the antibody activity in the IGM fraction. On the 10th day of the mani-fest rubella exanthema a bilateral hearing loss in the low frequencies sud-denly occurred. There was a complete recovery.

When the sudden hearing loss is a common symptom of the under-lying disease, as in the case of mumps and zoster oticus, there is no dif-ficulty in making a comprehensive diagnosis. If, on the one hand, a rare disease is occasionally associated with a sudden hearing loss, this may be a strong argument that both are related to each other. If, on the other hand, a sudden hearing loss is seen against the background of a very com-mon disease or a combination of common diseases, it is difficult to prove that these common diseases have anything to do with the sudden hearing loss. The etiological correlation between sudden hearing loss and con-comitant disease, highly probable with regard to the rare disease, less probable with regard to the common disease, may be demonstrated by 3 observations.

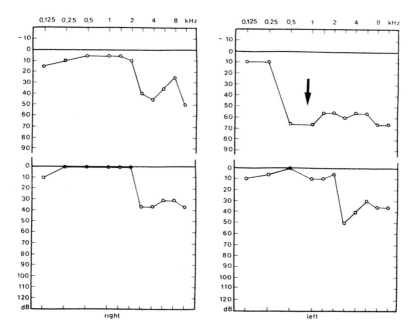

Fig. 28. Male patient of 53 suffering from hepatosplenomegaly with secondary porphyria, leukopenia and thrombopenia. Sudden hearing loss in his left ear. Full recovery.

A male patient of 37 (fig. 27) is known to have been suffering from Waldenström's disease (macroglobulinemia) since 3 years. His left ear shows a non-suppurative chronic otitis media. In his right ear he suffers a sudden hearing loss. The hearing ability deteriorates dramatically within the following 3 days. The blood picture showed extreme low values: Hb 4.8 g, rbc 0.6 mill., hematocrit 8%, leukocytes 2,400. Yet the hearing is restored fully. Repeatedly there have been reports of sudden deafness combined with Waldenström's disease. When one considers the extreme blood values a pathogenetic relationship between both is most likely.

A male patient of 53 is ill with a severe hepatosplenomegaly with a secondary porphyria, a leukopenia of 2,500 and a thrombopenia of 50,000. He suffers a sudden hearing loss (fig. 28). Fortunately the hearing is restored completely. In this case, too, a connection between basic disease and sudden deafness seems likely.

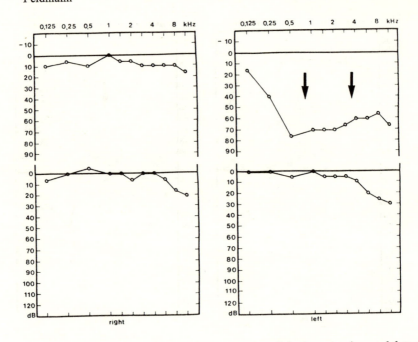

Fig. 29. Female patient of 52 suffering from a slight hypertension and latent diabetes. Sudden hearing loss in her left ear. Full recovery.

A female patient of 52 (fig. 29) had been suffering from severe headache in the left occipital region for a fortnight, which could not be cured by medicines. She was known to have a mild hypertension and a latent diabetes, which was treated by means of a diet. The sudden deafness happened without any extraordinary circumstances. By a treatment with Rheomacrodex® (Dextran 40, 10%) and Complamin® (nicotinic acid), the hearing was restored to normal and the headache disappeared completely. In this case the general situation and the basic diseases are so trivial and common that one hesitates to attach any importance to them with regard to the sudden deafness, for millions of people have the same constellation of basic disorders but do not suffer a sudden hearing loss. The fact, however, that not only the hearing loss but also part of the concomitant symptoms are favorably influenced by the treatment gives credit to the assumption that there is an underlying causal relationship between these two features.

Theories on Etiology

The etiology of the clinical feature 'sudden deafness' is certainly not uniform. This fact was demonstrated clearly by the last cases. In general the most important theories discussed are: viral infection; disturbed circulation, and lesion of round or oval window membrane.

Pathological evidence of ears which had suffered a sudden deafness is rare. Among others, 8 cases were published by *Schuknecht* et al. [21]. They all showed alterations in the cochlea similar to those observed after viral labyrinthitis, especially after mumps: atrophy of the organ of Corti, the tectorial membrane and the stria vascularis. On the other hand alterations that can be seen in animal experiments after the blocking of the blood supply, i. e. fibrous and osseous metaplasia, have not been demonstrated so far in human beings with regard to sudden deafness. Therefore the pathological evidence seems to point to the viral etiology.

From a clinical point of view, however, in the majority of cases the vascular pathogenesis appears more plausible. Confirmed cases of viral etiology of sudden deafness, e. g. following mumps or zoster oticus, have a decidedly unfavorable prognosis. On the other hand, they are not prone to relapse. As to the typical idiopathic sudden deafness there is often a good recovery and relapses are common.

Serologic evaluations with patients with sudden deafness aimed at a great number of viruses have confirmed a viral infection only in an insignificantly small number of cases, and even in these cases it is uncertain if the infection had caused the sudden deafness [14, 26]. In the overwhelming number of cases of sudden hearing loss it is impossible with the methods available at present to prove that they are associated with a viral infection.

The theory that a sudden hearing loss is often caused by a disturbed circulation has not been confirmed by pathological evidence, but it agrees with a number of clinical observations: suddenness of onset; frequently good recovery; apparently good response to treatment directed against circulatory disorders; liability to relapse; trigger function of fright, stress, shock and other vasoactive influences, and obvious connection with blood diseases such as leukemia, macroglobulinemia, etc.

There are quite a number of pathogenetic mechanisms which might be responsible for such circulatory disturbances: spasm of artery; arteriosclerosis; alterations of rheologic parameters; sludge phenomenon, and microembolization. These features might explain that there is no pronounced distribution in sudden deafness, relating to age.

Probably several factors must coincide to bring about the disturbances in the inner ear: anatomy of blood vessels; autonomic innervation, and metabolism, etc.

Some of these may be innate, some may be acquired; together they form a precarious situation in the inner ear waiting to be released by some banal incident. This could explain that only a few individuals suffer a sudden deafness but often do so recurrently, and it would explain that these individuals frequently have a hearing problem prior to the acute lesion.

The relationship between sudden deafness, fluctuating hearing loss and Menière's disease seems to be very close. In clinical practice sometimes these diagnoses cannot be differentiated. Cases which have started as a typical idiopathic sudden deafness later on reveal the specific symptoms of Menière's disease (fig. 21). Appearances of fluctuating hearing loss may as well be described as a special type of sudden hearing loss (fig. 7).

The pathology of Menière's disease, the hydrops of the endolymphatic system, is well known. In the idiopathic sudden deafness such a hydrops has not yet been demonstrated. It is doubtful if the hearing loss in the lower frequencies, which is very common in sudden deafness and typical in Menière's disease, may be interpreted as the functional correlate of this hydrops. For the time being such an assumption is no more than a working hypothesis.

Simmons [23] interprets the low-frequency loss as a manifestation of mechanical alterations in the cochlea and thinks of ruptures or tears in Reissner's membrane as being similar to those confirmed in Menière's disease.

Apparently the audiologic symptoms do not permit a clear inference as to the site of the lesion. They allow, however, a cautious prognosis. All statistics have shown that the low-frequency hearing loss has a much better prognosis than the high-frequency loss.

The theory that sudden deafness is caused by a rupture of the membrane of the round or oval window seems appropriate only to those rare cases in which the deafness has occurred under extreme external influences such as diving. We have no observations which would prove beyond reasonable doubt that a sudden hearing loss under normal environmental circumstances has been due to a rupture of the window membrane.

Uncertainty of Therapeutic Effect

I do not intend to discuss the innumerable suggestions that have been made with regard to therapy of sudden deafness. It is quite certain that there is a considerable rate of spontaneous remissions. *Mattox and Simmons* [13] calculate that 65–72 % of their 166 cases had a spontaneous remission. If this is correct, the treatment must yield a significant higher rate of success before it is to prove a therapeutic effect.

Null therapy or double-blind studies have not yet been carried out. The predominant hypothesis is: the prognosis of sudden deafness is most favorable if the treatment – here everybody of course means his own regimen – is started at the earliest moment. It becomes increasingly poor if the treatment is delayed. This hypothesis may be right, but it may just as well be a fallacy or a self-fulfilling prognosis: a treatment that is started early covers the spontaneous remission, a treatment that is started late is attempted because there has been no spontaneous remission.

We are still far away from a complete understanding of sudden hearing loss, but I hope that a careful study of individual cases rather than large statistics will reduce this phenomenon in the future more and more to a symptom which can be related to different diseases and thus becomes amenable to rational therapy.

Summary

After a brief statistic survey of general aspects the variety of clinical findings in sudden hearing loss is demonstrated by numerous observations. These illustrate the questionable role of supposed causes, the different modes of onset and course of sudden hearing loss, the audiometric findings and concomitant symptoms. The situation of both ears prior to the acute loss seems to be of great importance and offers a frame for classification, which is of clinical relevance These features and others are discussed with regard to common theories on the etiology of sudden hearing loss. The difficulties in assessing therapeutic effects are pointed out considering the high rate of spontaneous recovery.

References

1   Byl, F. M.: Seventy-six cases of presumed sudden hearing loss occurring in 1973: prognosis and incidence. Laryngoscope, St Louis *87:* 817–825 (1977).

2   Citelli, S.: Surdité rapide par simple congestion cochléaire. Oto-rhino-lar. int. *10:* 321–328 (1926).
3   Dishoeck, H. A. E. van; Biermann, T. A.: Sudden perceptive deafness and viral infection (report of the first hundred patients). Ann. Otol. *66:* 963–980 (1957).
4   Feldmann, H.: Das Refsum-Syndrom, klinische und audiologische Befunde einer eigenen Beobachtung. Archs Oto-Rhino-Lar. *227:* 379–382 (1980).
5   Fowler, E. P.: Sudden deafness. Ann. Oto-Rhinol. *59:* 980–987 (1950).
6   Jaffe, B. F.: Clinical studies in sudden deafness. Adv. Oto-Rhino-Laryng., vol. 20, pp. 221–228 (Karger, Basel 1973).
7   Kemper, J.: Der Hörsturz bei gesundem und vorgeschädigtem Ohr der Gegenseite; Inaugural-Diss., Münster (1977).
8   De Kleyn, A.: Sudden complete or partial loss of function of the octavus system in apparently normal persons. Acta oto-lar. *32:* 402–429 (1944).
9   Kreppel, H. G.: Beschleunigung der Blutkörperchensenkungsgeschwindigkeit (BSG) und Hörsturz. Lar. Rhinol. *58:* 954–961 (1979).
10 Kumpf, W.; Wandhöfer, A.: Zum Hörsturz bei vorgeschädigtem Innenohr der Gegenseite. Z. Lar. Rhinol. *51:* 838–841 (1972).
11 Lehnhardt, E.: Plötzliche Hörstörungen auf beiden Seiten gleichzeitig oder nacheinander aufgetreten. Z. Lar. Rhinol. *37:* 1–16 (1958).
12 Lehnhardt, E.: Der akute Hörsturz. Münch. med. Wschr. *102:* 2617–2621 (1960).
13 Mattox, D. E.; Simmons, F. B.: Natural history of sudden sensorineural hearing loss. Ann. Otol. *86:* 463–480 (1977).
14 Mercke, U.; Nordenfeldt, E.; Sjöholm, A.: Die Rolle einer Virusinfektion bei Hörsturz. HNO *28:* 125–127 (1980).
15 Morimitsu, T.: New theory and therapy of sudden deafness. Proc. Shambaugh 5th Int. Workshop on Middle Ear Microsurgery and Fluctuating Hearing Loss, Chicago 1976 (Strode, Huntsville 1977).
16 Neveling, R.: Die akute Ertaubung (Kölner Universitäts-Verlag, 1967).
17 Paparella, M. M.: Otological manifestations of viral disease. Adv. Oto-Rhino-Laryn., vol. 20, pp. 144–154 (Karger, Basel 1973).
18 Radü, H. J.: Hörsturz bei Kindern. Archs Oto-Rhino-Lar. *227:* 373–376 (1980).
19 Russolo, M.; Poli, P.: Acute idiopathic auditory failure: prognosis. Audiology *19:* 422–433 (1980).
20 Shaia, E. T.; Sheehy, J. L.: Sudden sensori-neural hearing impairment: a report of 1,220 cases. Laryngoscope, St Louis *86:* 389–398 (1976).
21 Schuknecht, H. F.; Kimura, R. S.; Naufal, P. M.: The pathology of sudden deafness. Acta oto-lar. *76:* 75–97 (1973).
22 Siegel, L. G.; Paparella, M. M.: Progress report: The National Registry for Idiopathic Sudden Deafness. Proc. Shambaugh 5th Int. Workshop on Middle Ear Microsurgery and Fluctuating Hearing Loss, Chicago 1976 (Strode, Huntsville 1977).
23 Simmons, F. B.: Theory of membrane breaks in sudden hearing loss. Archs Oto-lar. *88:* 41–48 (1968).
24 Simmons, F. B.: Sudden idiopathic sensori-neural hearing loss: some observations. Laryngoscope, St Louis *83:* 1221–1227 (1973).

25 Wandhöfer, A.: Der einseitige akute Hörverlust in Vergleich mit dem konseku-
    tiven Hörverlust des zweiten Ohres; Inaugural-Diss., Münster (1971).
26 Wilmes, E. H.; Roggendorf, M.: Zur Virusätiologie des Hörsturzes. Lar. Rhinol.
    58: 817–821 (1979).

Prof. Dr. H. Feldmann, ENT Clinic, University of Münster,
Kardinal von Galen Ring 10, D-4400 Münster (FRG)

Adv. Oto-Rhino-Laryng., vol. 27, pp. 70–82 (Karger, Basel 1981)

Sudden Deafness and Vertigo in Children and Juveniles

M. Berg, H. Pallasch

HNO-Universitätsklinik Erlangen, Erlangen, FRG

Introduction

Mainly vascular reasons are normally blamed for pathogenesis of sudden deafness with or without vertigo: for instance, a defect in alimentation during general hypotension, reduced circulation due to a narrowing of the respective afferent vessels in cases of arteriosclerosis together with their various disposing factors, maybe stenoses in the region of arteriae vertebralis and basilaris. As far as the causes for these vascular pathomechanisms are for instance arteriosclerosis or diabetes, they only exceptionally give a plausible etiology for sudden deafness in children and juveniles. No doubt possible reasons for vascular pathologies are a vasospasmic tendency, usually combined with several vegetative disorders, or systemic diseases like Cogan's syndrome or Waldenström's macroglobulinemia, tendency for fat embolisms or hypercoagulation, as well as the 'sludge' phenomenon triggered by shock. But all these reasons are supposed to be rare and exceptional too. More likely in children and juveniles are bacterial, viral or allergic inflammations of the inner ear, intoxications, maybe mechanical causes like ruptures of the round window or hemorrhage into the inner ear; surely there are also cases of psychogenic sudden deafness in children, and possibly tumors. Nevertheless, the question still arises whether there is an idiopathic sudden deafness, a so-called special children's sudden deafness, maybe due to disorders of the electrolyte composition of peri- or endolymph within the inner ear. The aim of the present study was to deliberate the pathogenic factors in children's and juveniles' sudden deafness in more detail.

Table I. Quotations from the DIMDI Index of Literature concerning 'sudden deafness in children or juveniles'

Year	Author	Cases	Causality
1976	*Alk*	1	rubella
1977	*Daniilidis* et al.	2	epidemic parotitis
1977	*Ishii and Toriyama*	1	neuronitis
1977	*Plentz*	2	psychogenic
1977	*Mair and Elverland*	1	vaccination impairment
1978	*Pashley and Shapiro*	1	spontaneous rupture of the round window
1979	*Mulch and Handrock*	1	intoxication by heroin

Table II. Quotations not contained in the DIMDI Index of Literature concerning 'sudden deafness in children or juveniles'

Year	Author	Cases	Causality
1952	*Wullstein*	1	focus
1960	*Weber*	2	focus
1962	*Saboulin*	1	unknown
1968	*Kessler*	1	stress disease

Overview of the Literature

Source papers concerning sudden deafness and vertigo in children and juventiles are rather rare: the 143 quotations stored at the DIMDI, the German Institute für Medical Documentation and Information in Köln, under the keyword 'sudden deafness', only contain 7 items explicitly related to 'children' or 'juveniles'. All of them are reports on single cases (table I). Some publications from the pre-DIMDI era are mentioned here without claim to completeness (table II). They also contain reports on single cases. It is striking that always a manifest cause or reason for the deafness was found – with only one exception; a 'real' sudden deafness in the sense of a disease sui generis in children and juveniles does not seem to exist.

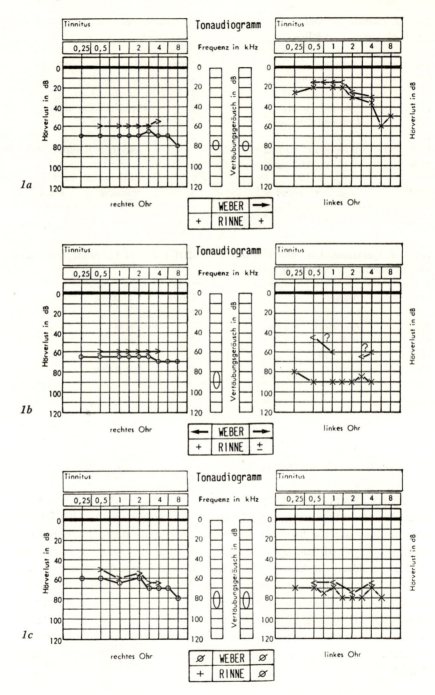

1a

1b

1c

If it is true that an essential reason, disturbed microcirculation, can only seldom or never be blamed for sudden deafness up to the age of 20 years, then: (a) sudden deafness within this age group must be much more rare, and (b) it must be easier to find and recognize sudden deafness of a different etiology, because they are no longer hidden within the large number of cases of vascular origin.

Clinical Overview

During the last 3 years 17 children and juveniles, the age below 21, came to our hospital and got the diagnose of a sudden deafness. Generally the following examinations took place: Anamnesis and complete ENT status; tone and speech audiogram, suprathreshold audiometry when necessary and possible, by part middle ear impedance measurements, and in part objective audiometry; a comprehensive vestibular examination; informative clinical laboratory testing; X-rays of the maxillary sinuses and the ears, normal and computer-aided tomography when especially indicated for exclusion of a tumor. The pediatric hospital also performed consultative examinations with special attention on system diseases and neurological pathologies. In the 8 patients of the last year several virus examinations were performed.

Case Reports

We present 2 of our 17 patients as exemplary cases:

K. D., 19 years, female, had been under a specialist's treatment from the age of 4 because of impaired hearing on her right ear (fig. 1a). The first audiogram, taken at the age of 12, showed an omnifrequential hearing loss of 60 dB in the right ear and a moderate high frequency loss in the left. At the age of 19 years she came

Fig 1. K. D., 19 years, sudden deafness in the left ear. *a* An audiogram taken at the age of 12 displays an omnifrequential sudden hearing loss on the right, and a moderate high-frequency loss on the left. *b* Audiogram just after incidence of the sudden deafness, and *c* 6 months later. The left ear did not respond to therapy and shows now about the same hearing loss as the right.

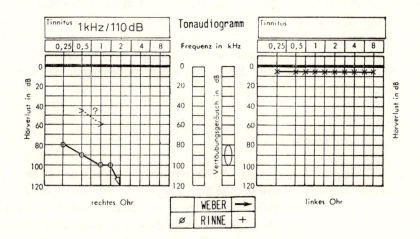

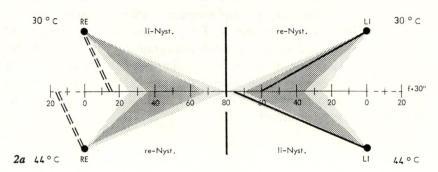

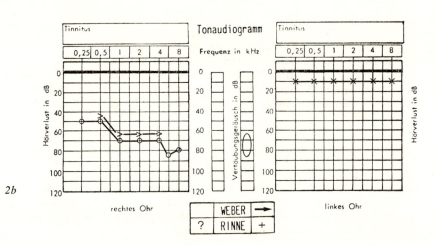

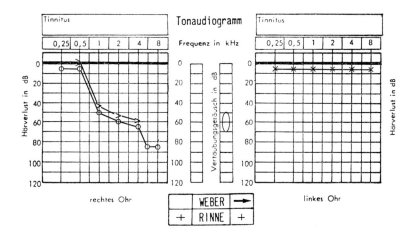

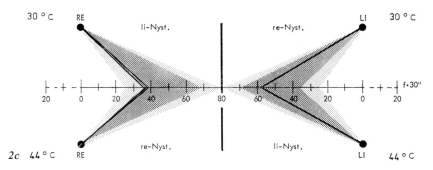

Fig. 2. M. M., 13 years, sudden deafness in the right ear. *a* Audiogram of the severe hearing loss. The calorigram indicates no sensitivity of the right labyrinth, only a left beating spontaneous nystagmus. *b* Audiogram after a 10-day infusion therapy. *c* Audiogram taken 3 months later indicates restored hearing for low frequencies, the calorigram shows restored function of the right labyrinth.

to our hospital because the hearing ability on her good left ear had suddenly decreased. She had experienced this together with some tinnitus 2 weeks before; later she had had an infection accompanied by highly increased temperature. The ENT inspection was nearly normal. X-Rays of the maxillary sinuses and of the ears were without pathological results. The audiogram showed the old threshold on the right ear, but on the left ear the threshold now was at 80–90 dB (fig. 1b). The neurotological examination yielded no spontaneous, but a positioning and head-shaking nystagmus, and a gaze nystagmus. The caloric stimulation of the left ear provoked only a reduced nystagmus reaction; so there were signs of central as well as of peripheral disorder. The examination by a neurological consultant showed no pathological result. Because of the fever infection some viral examinations were performed in which the value for influenza A was increased by a factor of 128:1; the other viral test results concerning adenovirus, herpes simplex virus, varicella-zoster virus, cytomegalovirus, influenza B, parainfluenza 1–3 and choriomeningitis were normal. The enhanced value for influenza A was renormalized in a later follow-up. A computer tomography was without signs of a cerebello-pontine angle tumor, which by zisterno-meatography could be definitly excluded. The patient was not responsive to the infusion therapy. The last control audiogram half a year after discharge from hospital showed in both ears an omnifrequential hearing loss of 60–80 dB (fig. 1c).

Another patient, 13-year-old male M. M., experienced a sudden deafness and tinnitus in his right ear together with vertigo and nausea after he had had a strong cold infection. The ENT inspection was normal except for large fissured tonsillae. The audiogram showed normal hearing on the left and a 80- to 100-dB hearing loss on the right ear (fig. 2a). The vestibular examination yielded a direction-fixed spontaneous nystagmus beating to the left, a positioning nystagmus and no reaction to caloric stimuli of the right ear. The BSR (blood sedimentation rate) was clearly increased to 28/55 mm. All virological examinations were without pathological results. A pediatric examination gave no hint to a reason of the hearing loss. The neurological examination was without pathological findings. After infusion treatment the tone threshold improved up to 50–70 dB (fig. 2b). A control audiogram, 3 months later, showed complete restoration of hearing for the low frequencies up to 1,000 Hz, but still a steep slope down to 80 dB for the high frequencies (fig. 2c) A vestibular control examination gave no pathological results, except a slightly reduced reaction to caloric stimuli in the right ear. The only hints for a possible pathomechanism in this case are the cold infection immediately before the sudden deafness and the increased BSR.

Results and Discussion

Out of 329 patients suffering from sudden deafness and treated in our hospital since August 1977, only 17 (i. e. 5%) had an age below 21 years (fig. 3a). Within the normal population about 30% belong to this group. In other words: About one third of the population share only a twentieth of the cases of sudden deafness (fig. 3b). This means that for children and juveniles an etiology which is frequent elsewhere is not applicable. *Shaia and Sheehy* [1976] report in an overview of over 1,220 cases 17 with an age below 10 years and 159 below 30 years. This corresponds fairly well with our age distribution, having 16% below 30 years. The youngest of our patients had an age of 9 years and we must remark that so-called 'sudden deafness' after complications in the middle ear were not counted in our investigation.

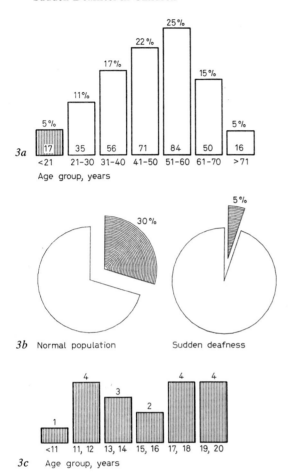

3a

Age group, years

3b   Normal population   Sudden deafness

3c   Age group, years

Fig. 3. 'Statistics' of sudden deafness: *a* Distribution of sudden deafness over age: Since August 1977 a total number of 329 patients were treated in Erlangen. *b* The youngest 30% of the population share only 5% of the cases of sudden deafness. Therefore, a frequent etiology, for instance vascular problems, does not apply for children and juveniles. *c* All teenagers have about the same risk of suffering sudden deafness.

The distribution versus age and sex shows the same incidence for all 'teenagers', but two thirds female and only one third male patients (fig. 3c). (Due to the small total number of cases of only 17, all the following statements must be understood in the sense of a trend deliberation not claiming to be statistically perfect.) In adults the female:male ratio is re-

Table III. Hearing ability before sudden deafness

	Ipsilateral ear	
	normal	disturbed
Contralateral ear		
Normal	9/17	1/17
Disturbed	1/17	6/17

versed: 6 women suffering from sudden deafness correspond to 10 men out of all our adult sudden deafness cases. This can possibly be understood, considering the increased tendency to pathologies of the vessel system which males are supposed to have. At the same time one can also conclude that in fact sudden deafness due to vascular reasons is more rare for young patients.

The hearing losses of our young patients as measured for the first time after the acute event were between 30 dB and complete deafness. The small number of only 17 cases does not allow one to show mean audiograms or only typical ones. There were high-frequency losses (5 cases), omnifrequential (5 cases) and low-frequency losses (3 cases) and 4 cases of complete or nearly complete deafness.

For the time preceding the sudden deafness about the half of the patients remembered a normal hearing, about one third reported an binaural impairment; in part the latter patients had already been treated by specialists. Only one of our 17 patients had already been completely deaf in the other ear (table III).

Tinnitus and vertigo used to coincide: 6 of 15 patients suffered from both, another 6 from neither (table IV). Only 1 showed vestibular pathologies but no tinnitus, only 2 reported a tinnitus and were free of vestibular disease. We have not found any young patients suffering only from acute vertigo but without hearing loss.

As to the laboratory findings, the BSR was increased only in the above presented case. Leukocytes were increased slightly in 2 cases, strongly in 1 case (the latter was a 9-year-old boy, who in consequence of a later proven obstruction of a. carotis showed a temporary facial palsy of central type and a hemiplegia beside his sudden hearing impairment). All fat values of the blood were normal when determined. Only 1 female

Table IV. Tinnitus and vestibular disorder

	Tinnitus	
	no	yes
Vestibular disorder		
No	6/15	2/15
Yes	1/15	6/15

Table V. Success of therapy versus prior sudden hearing loss

	Therapy successful	
	yes	no
Prior sudden hearing loss		
No	5/14	3/14
Yes	1/14	5/14

patient showed hypotension (RR = 90/50); she suffered from recurring low-frequency hearing losses which responded well to infusion therapy.

After infusion therapy, which is applied with certain slight changes using dextran, cortisone, antibiotics and novocain in our hospital, only 6 patients of 16 showed clear improvement, i. e. part restoration of normal hearing. Unfortunately in 2 cases the hearing ability worsened during therapy, in 6 of 16 cases it stayed constant (1 patient had refused the therapy). It does not seem possible to give even a tendential relation between type and degree of the hearing loss and success of therapy.

There is a certain trend insofar as young patients with an already impaired hearing before the acute hearing loss do not respond well to infusion therapy; but the others with good hearing before are not much more successful (table V). Maybe one must ask how many of the former are of the type 'hereditary degenerative deafness', but if the sudden deafness is seen as a case of ENT emergency this question should not arise before treatment.

Table VI. Success versus vestibular involvement

	Therapy successful	
	yes	no
Vestibular disorder		
No	3/15	4/15
Yes	4/15	4/15

From our small collective it is not possible to confirm the general experience that the prognosis of sudden deafness is better in cases without vertigo than in cases including vestibular symptoms: there is no tendency at all (table VI).

Table VII gives a view on the probable reasons for the sudden deafness and on the success of the therapy in our young patients. 2 cases seem to us to be of purely psychogenic origin, 1 case both of hereditary and psychogenic origin. Only 2 cases can be derived from a virus infection, namely influenza A. 1 case was clearly of vascular origin, 1 case probably triggered by a rubella vaccination. After exclusion of all other thinkable causes we can call 4 cases 'idiopathic'. In 1 case we now believe that the hearing impairment had existed for a long time and is of unknown origin. For the remaining 5 cases it is not possible to give reason, partly because the patients are no longer at our disposition, partly because not all diagnostic possibilities have been used.

Conclusions

(1) From our not very numerous cases of sudden deafness in young patients we can nevertheless draw the conclusion that a unique clinical picture does not exist. (2) Contrary to sudden deafness in adults vascular reasons do not play an important role. (3) Viral infections are not of the importance we originally had expected. (4) Young patients, who had had a sensineural hearing impairment already before the sudden deafness, seem to have a bad prognosis. (5) Vestibular involvement seems to be of no meaning as to the success of therapy to be expected.

Table VII. Probable etiology of sudden deafness children and juveniles below 21 years: 17 cases since August 1977

Patient	Age, years	Sex	Probable etiology	Comments and results
L. K.	18	F	psychogenic	ERA and impedance normal, subjectively profound deafness
H. V.	14	M	psychogenic	therapy without success, normal hearing after training
J. U.	12	M	semi-psychogenic	fluctuating hearing loss, good success
K. M.	13	F	influenza A	deteriorated hearing after infusion therapy
K. D.	18	F	influenza A	unsuccessful treatment
D. B.	9	M	a. carotis	hemiplegia, spontaneous improvement of hearing
U. I.	14	F	rubella vaccination	completely deaf
M. M.	12	M	idiopathic	good success
J. S.	16	M	idiopathic	no change
H. G.	19	F	idiopathic	recurrent hearing loss, successful therapy
M. I.	11	F	unknown	profound deafness for many years; no change
H. H.	17	M	unknown	good success, no complete diagnostic
P. H.	17	F	unknown	refused treatment
H. K.	15	F	still unknown	just under treatment
M. M.	20	F	still unknown	facial palsy, vestibular central disorder, hearing improved
S. Z.	20	F	unknown	noise, stress, still under observation

Summary

The 17 cases of sudden deafness in children and juveniles treated in the HNO-Universitäts-Klinik Erlangen between August 1977 and December 1980 are analyzed and correlations of symptoms and possible etiologies are discussed. There is no unique clinical picture. Vascular reasons do not play an important role contrary to adult patients. Unexpectedly viral infections are not of great importance. Young patients who had a sensineural hearing loss prior to the sudden deafness seem to have a bad prognosis for therapy. Vestibular involvement seems to be of no meaning for a successful therapy in children and juveniles.

References

Alk, G. P.: Einseitiger reversibler Hörverlust nach Rubeolen beim Erwachsenen. Lar. Rhinol. Otol. *54:* 379–384 (1976).

Daniilidis, J.; Petropoulos, P.; Iliadis, T.: Sudden deafness in connection with mumps. Lar. Rhinol. Otol. *56:* 432–435 (1977).

Ishii, T.; Toriyama, M.: Sudden deafness with severe loss of cochlear neurons. Ann. Otol. Rhinol. Lar. *86:* 541–547 (1977).

Kessler, L.: Akuter Hörsturz im Kindesalter. HNO *16:* 148–149 (1968).

Mair, I. W.; Elverland, H. H.: Sudden deafness and vaccination. J. Lar. Otol. *91:* 323–329 (1977).

Mulch, G.; Handrock, M.: Sudden binaural deafness after acute heroin intoxication. Lar. Rhinol. Otol. *58:* 435–437 (1979).

Pashley, N. R.; Shapiro, R.: Spontaneous perilymphatic fistula. J. Otolar. *7:* 110–118 (1978).

Plentz, R. J.: Juvenile psychogenic hearing disorders. Fortschr. Med. *94:* 989–995 (1977).

Saboulin, D.: Surdité de perception passagère chez un enfant. J. fr. Otorhinolar. *12:* 372 (1962).

Shaia, F. T.; Sheehy, J. L.: Sudden sensori-neural hearing impairment: a report on 1,220 cases. Laryngoscope, St Louis *86:* 389–398 (1976).

Weber, I.: Ein Beitrag zu den kryptogenen Hörstörungen des Innenohres. Lar. Rhinol. Otol. *39:* 589–595 (1960).

Wullstein, H.: Schwerhörigkeit in Abhängigkeit von einem Focus? Lar. Rhinol. Otol. *31:* 541–545 (1952).

M. Berg, PhD, HNO-Universitätsklinik, Waldstrasse 1, D-8520 Erlangen (FRG)

Adv. Oto-Rhino-Laryng., vol. 27, pp. 83–99 (Karger, Basel 1981)

Electrophysiological Findings in Patients with Sudden Deafness: A Survey

M. Hoke, B. Lütkenhöner

Experimental Audiology, Ear, Nose and Throat Clinic,
University of Münster, Münster FRG

Introduction

An examination of publications related to the sudden loss of the cochlear function gives the impression that the term 'sudden hearing loss' (or 'sudden deafness') is more frequently used as a heuristic description of a clinical picture than a designation of a disease sui generis. This impression is not surprising when one thinks of the considerable difficulty of establishing a fairly unchallengeable definition of the disorder in question. *Feldmann* [1981], who provides a thourough analysis of sudden hearing loss, stresses five criteria which characterize this doubtless multifactorial event and which may serve as a heuristic definition: Unilateral occurrence, sudden onset, obviously sensory origin, the lack of an identifiable exogenous cause, and the lack of a discernible connection to other diseases.

With respect to diagnosis and therapy, the questionable etiology and uncertain pathogenesis of sudden deafness is highly unsatisfactory. While morphological evidence of ears which had suffered a sudden hearing loss is rare anyhow, electrophysiological findings are exceptionally few. This is most likely due to doubts about the exposure of an already impaired ear to higher sound intensities. The difficulties of electrophysiological research are further enlarged by the fact that a sufficiently reliable experimental model of sudden hearing loss does not exist so far.

When we look for possibilities of gaining some more insight by means of objective methods like electric response audiometry (ERA), we first need to agree about some basic assumptions with regard to sudden hearing loss. Despite the uncertainty of its etiology it is a widely accepted working hypothesis that genuine sudden hearing loss (in this context only the genuine sudden loss of the cochlear function is considered; cases of sudden hearing loss with retrocochlear origin are excluded) is caused by a cochlear impairment (most likely a hair cell impairment and, ipso facto, a subsequent degeneration of the primary auditory nerve fibres). Consequently, we ought to expect signs of recruitment (i. e. an abnormally rapid increase in the loudness sensation along with elevated threshold), and the corresponding findings in objective measures revealed by electrophysiological methods. This implies that the only procedures of ERA which are valuable are those which record potentials of cochlear origin (electrocochleography: compound cochlear microphonics (the term 'compound cochlear microphonics' has been coined in analogy to the term 'compound action potential' as it comprises the sum of the output of a large number of individual generators [*Hoke and Hieke*, 1976; *Köpcke* et al., 1980]), summating potential, and compound action potential of the auditory nerve, because these potentials alone would not only reflect any hearing impairment but would be sufficient to substantiate its cochlear genesis. Brainstem and cortical evoked response audiometry are disregarded in this context because they do not reveal any significant information about the early cochlear potentials. This implies as well that – similar to the situation in psychoacoustic audiometry – we would not be able to distinguish between sudden hearing loss and other hearing disorders of likewise suspected (Menière's disease, fluctuating hearing loss) and proven (ototoxic, noise-induced) cochlear etiology. The differential diagnosis, again, would turn out to be a diagnostic procedure per exclusionem.

In collecting the material on which this paper is based, the pertinent literature was scrutinized and, in addition, a request for all available electrophysiological data and publications related to sudden hearing loss was circulated among all members of both the 'International Electric Response Audiometry Study Group' (IERASG) and the 'Arbeitsgemeinschaft Deutscher Audiologen und Neurootologen' (ADANO). Despite the broad response, the outcome of the inquiry was exceptionally meagre. Just some 3 % of the inquired persons were able to submit more or less relevant information. Results of electrophysiological investigations which give more detailed insight into the pathologic processes taking place within the cochlea can be counted on one's fingers. Nevertheless, the authors would like to use this opportunity to express their gratitude to all those who responded to the circular even if they were unable to contribute any findings of their own. The result obviously demonstrates doubts existing among audiologists as to whether patients with supposed sudden hearing loss should be subjected to extensive audiometric investigations because the inevitable sound exposure might act as an additionally injurious factor. *Mair and Laukli* [personal communication, 1979], for example, definitely responded that their '. . . departmental policy with respect to sudden deafness is to avoid possibly nocuous investigations, including supra-threshold audiometry, in the acute stages. We therefore do not perform electric response audiometry in these cases.'

Physiological Foundations and Normal Findings

Before considering the pathological findings in patients suffering from sudden hearing loss of cochlear origin, some physiological foundations will be outlined briefly as far as they are relevant for the interpretation of electrocochleographic data.

As already mentioned, three different compound potentials can be recorded with electrocochleographic techniques: one potential of neural origin, the compound action potential of the auditory nerve (AP), and two receptor potentials, the compound cochlear microphonics (CM) as well as the summating potential (SP).

Compound Cochlear Microphonics

It seems to be sufficiently established now that – as a result of the transduction process taking place in the sensory cells – the intracellular recorded receptor potential is the electrical analogon of the mechanical stimulus effective on the receptive structures of the hair cells [*Sellick and Russell,* 1978]. Its AC component (the intracellular equivalent of cochlear microphonics) exhibits – like the DC component (see below) – the same frequency selectivity [e. g. shape of tuning curves and lability of tuning: *Sellick and Russell,* 1978] and shows the same nonlinear features [e. g. two-tone suppression: *Sellick and Russell,* 1979]. The so-called 'second filter' [*Goldstein,* 1967; *Evans,* 1972] which has been proposed to account for the much sharper tuning of auditory nerve fibres as compared to the relatively moderate tuning of the basilar membrane, is considered now to be more peripherally located than was originally assumed. Hence, specified features and their pathological alterations as found in the discharge pattern of single auditory nerve fibres do not necessarily reveal special properties and alterations of the auditory nerve fibres. They may also demonstrate properties belonging to the transduction process in the hair cells, or even exhibit disturbances in the coupling between the receptive field of the sensory cells and the surrounding vibrating structures of the cochlea ['cochlear micromechanics': *Allen,* 1977].

The statements made up until this point only apply to single hair cell data. Though single cell recordings are not possible in electrocochleography, this digression into some special aspects of stimulus transduction as well as localization of frequency selectivity in the cochlea appears to be of great importance as several arguments in the following paragraphs and chapters will demonstrate. It should also remind us that we ought to be aware of the possibility that pathological electro-cochleographic findings in specified parameters of the compound AP must not nec-essarily be of (exclusively) neural origin.

In contrast to the importance which is attached to single hair cell recordings, the significance of compound CM for clinical diagnosis is doubtful [*Eggermont,* 1976b; *Hoke,* 1976b]. This limitation is caused by several different factors:

Electroanatomy and Mechanics of the Cochlea. The compound CM is the weighted average [*Kohllöffel,* 1971] of a large number of receptor potentials (of individual hair cells) with different amplitude and phase values [*Dallos,* 1973; *Hoke,* 1976b; *Hoke and Hieke,* 1976]. Since the attenuation of the individual potentials increases with increasing distance between the locus of origin and the recording site, the round window [*von Békésy,* 1951; *Honrubia* et al., 1976], the recording electrode essentially 'sees' potentials originating from the basal third of the cochlea so that the transfer function of the compound CM mainly assumes the shape of the com-bined transfer functions of the outer and middle ear [*Hieke and Hoke,* 1976; *Hoke,* 1976b].

Recording Site. The usual transtympanic placement of the electrode without visual control adds further uncertainties due to the influence of the recording site on the frequency response curve [*Köpcke* et al., 1980].

Artifacts. Electromagnetical and resistive leakages as well as a microphonic effect of the electrode itself are not distinguishable from CM and can only be avoided by a most carefully designed experimental setup and an experienced in-vestigator (*Hoke,* 1976b].

Summating Potential

The SP represents a perstimulatory DC shift which is generated near the cuti-cular lamina of the hair cells either by a nonlinear vibration of the basilar mem-brane around its resting position, or a nonlinearity in the mechano-electric trans-duction process itself [*Whitfield and Ross,* 1965; *Durrant and Gans,* 1975], and therefore assumes the shape of the envelope of the stimulus signal [for more details, see *Eggermont,* 1976b]. The intracellular equivalent of the SP is the DC component of the receptor potential as observed by *Sellick and Russell* [1978]. It shows – similar to the AC component – the same frequency selectivity as the primary auditory nerve fibres [*Sellick and Russell,* 1978].

Recorded from the scala tympani relative to the neck muscles, the (compound) SP exhibits a tonotopic relationship to the place of maximum basilar membrane dis-placement. In the guinea pig it is found to be positive only in stimulated parts of the cochlea, whereas a negative SP is present in the parts which are at rest [*Kupper-man,* 1970]. The conditions are very similar when the SP is recorded from the round window or its vicinity. A positive SP only occurs when the ear is exposed to a high frequency stimulus.

In man, possibly due to a different electroanatomy [*Eggermont,* 1976c], the normal polarity of the round window- or promontory-recorded SP is negative, in-cluding high frequencies and high stimulus intensities, whereas a positive SP is the exception. Similar to compound CM recordings, the amplitudes of SP also show a considerable scatter [*Eggermont, 1976b*]. Its clinical use, however, seems to be significantly superior to that of CM.

Compound Action Potential

The compound AP of the auditory nerve represents the summed electrical activity of a large number of individual, synchronously stimulated auditory nerve fibres, as recorded with a remote electrode [for more details, see *Eggermont, 1976b; Eggermont* et al., 1974].

Waveform. The compound AP of the auditory nerve commonly assumes a diphasic shape. Its waveform is mainly determined by the time domain waveform of the stimulus signal, its spectral composition, and its intensity; it is further influenced by cochlear pathologies. Contrary to *Aran* et al. [1971] and *Aran and Negrevergne* [1973] who proposed a classification of pathological waveforms (recruiting, dissociated, broad and abnormal), it is well established now that a simple description of the compound AP waveform cannot be used as a diagnostic tool [*Eggermont, 1976b*].

Amplitude-Intensity Relationship. Drawn in a linear scale, the amplitude-intensity (I/O) curve is discontinuous. It consists of two sections, a slightly rising section below 60 dB HL [the low or L-part of *Yoshie, 1968*] and a steep rising section (H-part). This subdivision of the I/O curve into two parts of different slope has been attributed to two differently responding receptor-nerve fibre populations [*Yoshie, 1968*]. This two-population hypothesis had been the prevailing interpretation until *Evans* [1975] gave a more plausible explanation which was suggested by the experimental findings of *Kiang* et al. [1970]. His hypothesis to explain the recruitment phenomenon is based on one auditory nerve fibre population (innervating the inner hair cells) only. *Eggermont* [1977a] published an extensive collection of electrocochleographic data recorded in recruiting ears, which strongly support *Evans* [1975] one-population hypothesis (more details are reported below in the section on the 'recruitment phenomenon').

Latency-Intensity Relationship. The latency-intensity curves of tone-burst elicited APs normally show – in a logarithmic plot – a linear dependence on SPL. Only latency-intensity curves of wide-band, click-evoked APs exhibit a transition at about 60 dB HL [*Eggermont, 1976b*].

Frequency Specificity. Unlike click-evoked compound APs which only exhibit information from the most basal part of the cochlea, tone-burst evoked compound APs show a sufficiently high frequency specificity to determine the corresponding pure tone thresholds [*Eggermont* et al., 1974]. Supra-threshold findings derived from tone-burst evoked compound APs, however, must be interpreted with care (see also response areas, narrow-band AP).

Narrow-Band AP. Reliable frequency-specific information at supra-threshold stimulus intensities can be obtained only by using more sophisticated techniques. *Teas* et al. [1962] introduced the high-pass noise masking technique to derive narrow-band APs which originate from circumscribed portions of the cochlear partition ('derived APs'). This is done by subtracting click-evoked compound APs selectively masked with high-pass noise of two different lower cut-off frequencies. This technique has been successfully applied to investigations in man by *Eggermont* et al. [1974] and *Elberling* [1974].

Response Areas, Tuning Curves. If the high-pass noise masking technique is applied to tone-burst evoked compound APs, response areas as well as tuning curves can be obtained [*Eggermont*, 1976d, 1977b]. [It should be mentioned that, in his early publications, *Eggermont* uses the term 'response area', unlike the common practice in neurophysiology, for isoresponse contours (= tuning curves), whereas he refers to isointensity contours (= response areas) as amplitude distribution.] The shape of the tuning curves is very similar to that of tuning curves of auditory nerve fibres as well as to psychoacoustic tuning curves. This technique enables us to relate electrocochleographic findings in human to results obtained from single-fibre studies in the normal and pathological animal cochlea.

Experimental Models of Sudden Hearing Loss

Experimental models of a disease or disorder are extremely valuable in studying the underlying processes of this disease and in validating diagnostic procedures and therapeutic concepts. However, as pointed out in the introduction, a generally accepted experimental model of sudden hearing loss does not so far exist though several attempts have been made. *Jung and Rosskopf* [1975] referred to a 'hearing loss induced by furosemide' (which can also be induced by other ototoxic loop-inhibiting diuretics, e. g. ethacrynic acid or bumentanide). *Handrock* [1978] proposed the cryogenic damage of the cochlea as a model of an experimentally induced, acute hearing loss. In the authors' opinion, the cryogenic damage is superior to noise exposure. The effect to temporary hypoxia [*Evans,* 1974] also causes, at the cochlear level, effects like detuning of primary auditory neurons, which are common also in cases of sudden hearing loss. According to the definition of sudden hearing loss as given before (lacking identifiable exogenous cause, but probably – temporary – circulatory insufficiency), only the cryogenic hearing loss or that induced by a hypoxia would approximately fulfill these requirements.

Electrophysiologic Aspect of the Recruitment Phenomenon

The term 'recruitment' as coined by *Fowler* [1928] describes the (psychoacoustic) phenomenon that there is an abnormally rapid increase in loudness sensation with increasing stimulus intensities, along with an elevated threshold, ending up in a perception of equal loudness at high stimulus intensities. This phenomenon had originally been attributed to cochlear or nerve fibre pathology, later restricted to cochlear lesions, but is now more recently being questioned once again [*Eggermont,* 1977a].

The morphological finding of a complete loss of the (outer hair cells innervating) spiral nerve fibre population, along with the (inner hair cells innervating) radial nerve fibre population remaining intact, gave rise to the two-population hypothesis of loudness recruitment. This explanation of recruitment, however, turned out to be inconsistent with findings in Menière's disease, in which recruitment is a common finding in all stages, though the outer hair cells are quite or almost intact [*Lindsay,* 1968; *Schuknecht,* 1968, 1974]. *Kiang* et al. [1970] raised questions about the two-population concept. They demonstrated in ototoxically damaged ears that recruitment arises not from the selective loss of low-threshold nerve fibres but from the conversion into units with high thresholds at the characteristic frequency along with normal low-frequency behavior. This concept has been taken up by *Evans* [1975]. He concluded that in situations of cochlear deafness, the nerve fibres exhibit broad tuning curves; then, once the (elevated) threshold is reached, the number of stimulated units will increase relatively rapidly with stimulus intensity.

Electrocochleography and Recruitment

It is a well-known fact that in particular clinical investigators from early on tried to relate specified abnormal electrocochleographic findings to the psychoacoustically detected recruitment phenomenon. For that purpose, various features and parameters have been tried with little or no success (description of the waveform, slopes of input-output curves, and latency-intensity relationships of the compound AP). *Yoshie* [1968] interpreted the two differently rising parts of the I/O curve as a confirmation of the two-population concept, and he regarded the lack of the L-part as an indication of recruitment. The most thorough electrocochleographic study in recruiting ears was done by *Eggermont* [1977a]. He demonstrated that: (1) the shape of the compound AP depends more on the type of audiogram and the type of cochlear dysfunction than on the presence of recruitment [*Eggermont,* 1976b]; (2) conclusions about the presence of

recruitment based on the slope of I/O curves will be wrong in about 20 %
of the cases, and (3) the latency criterion for the presence of recruitment
is in fact a threshold criterion and should therefore be used with care.

A broadening of the tuning curve, however, seems to be connected
with the presence of recruitment for a particular tone-burst frequency.
These findings also support the one-population hypothesis for recruitment.
In another study carried out on 37 patients suffering from Menière's disease, *Eggermont* [1979a] demonstrated that detuning is related to a significant decrease of latency in the narrow-band AP. If this relationship
can be established in animal experiments, the measurement of narrow-
band AP latencies would become a valuable diagnostic means in electrocochleography.

Pathological Findings

Compound Cochlear Microphonics

It has been deduced in the section on the physiological foundations
that the clinical use of CM is only limited [*Hoke,* 1976b; *Eggermont,*
1976b]. Most investigators use the absence or presence of cochlear microphonics along with the presence or absence of the compound AP of the
auditory nerve merely to classify the type of hearing loss with regard to its
supposed origin: conductive, cochlear, retrocochlear [*Ruben,* 1967; *Beagley,* 1974; *Aran and Charlet de Sauvage,* 1976], or even doubt the validity
of compound CM recordings [*Eggermont* et al., 1974; *Aran and Charlet
de Sauvage,* 1976; *Eggermont,* 1976a, b; *Hoke,* 1976b]. *Ruben* [1967]
was the first to publish electrocochleographic results from a large number
of cases. The collection comprises 66 cases of profound, nonconductive
deafness. The first group, defined as cochlear deafness (no cochlear microphonics and compound APs of the eighth nerve present), consists of 22
patients. The etiology of every second case is stated to be unknown; there
are no indications of a suspected sudden hearing loss. *Naunton and Zerlin*
[1976] and *Zerlin and Naunton* [1978] report 2 cases of a severe sudden
hearing loss where normal hearing was recovered. The initially reduced
CM amplitudes during the period of depressed hearing showed an increase
during the course of recovery, and the authors conclude that hair cell
function was included in the sudden hearing loss. A case of Menière's
disease reported in the latter publication again shows reductions in CM
amplitude concomitant to a corresponding elevation of conventional pure

tone audiometric thresholds. The authors conclude that measurable changes in CM amplitude together with changes in the state of hearing indicate that hair cell activity is related to changes in hearing sensitivity. *Beagley and Gibson* [1978] mention the application of electrocochleography, i. e. the registration of compound CM and AP for the differential diagnosis of cochlear and retrocochlear lesions including sudden hearing loss and Menière's disease. Findings obtained in the first group are not reported, while CM in cases of Menière's disease tends to be small in amplitude and distorted.

Graham et al. [1978] and *Graham* [1979] published data collected in 100 cases of sudden hearing loss. Owing to a different definition, the authors included all cases of hearing loss with sudden onset irrespective of its origin. CM was used only to distinguish between cochlear and retrocochlear lesions. One case of a patient is included whose hearing completely recovered after the sudden onset of nearly one-sided deafness, in which electrocochleography showed 5 days after the onset of deafness a large CM, but no detectable AP.

Summating Potential

In contrast to compound CM, more information is available on the second receptor potential, the SP. Since the SP behaves differently in different hearing losses of supposed cochlear origin with recruitment present, findings in Menière's disease and hair cell loss of different origin are also to be reviewed. *Eggermont* [1976a–c, 1979b] demonstrated that the magnitude of the SP is reduced (or even not detectable) in cases of a loss of outer hair cells, whereas the SP magnitude is increased in cases of Menière's disease. The lower detection threshold for SP tentatively points to an enlarged asymmetry in the motion of the basilar membrane as well as to a static displacement of the basilar membrane already present at lower stimulus intensities, due to an endolymphatic hydrops. These findings have been confirmed by *Beagley and Gibson* [1978], *Moffat* [1979], *Moffat* et al. [1978], *Morrison* et al. [1980], *Rietema* [1979] and *Gibson* [1980]. According to *Rietema* [1979], the SP can be regarded as the only reliable parameter for estimating the functional state of (outer) hair cells in the basal turn of the cochlea. In Menière's ears, a positive SP or the absence of SP points to a change from a fluctuating to a more stable state of the disorder. According to *Eggermont* [1976a], a transition from negative to positive SP occurs at a frequency at which a threshold elevation also in the pure tone audiograms begins.

Nishida et al. [1976] reported about 34 cases of sudden hearing loss which were subjected to electrocochleographic assessment. A group of 15 patients showed complete recovery or remarkable improvement, while the remaining 19 patients showed just a slight improvement or even no change of hearing. The time elapsed until the first examination was 5.2 days on average in the first group, without a noticeable difference between two subgroups, 8 cases with a dominant –SP, and 7 cases with a high AP response as prevailing feature. The time elapsed until recovery or end of improvement amounted to 27.4 days on average in the first subgroup (dominant –SP), but to 15.4 days in the second one. The second group with no or almost no recovery showed at the first examination (12 days on average after onset) either positive or negative SP of normal or reduced magnitude, or no SP response at all. The authors regarded these findings as unfavorable prognostic criteria, whereas they tentatively suggested a dominant –SP (i. e. exceeding the N_1 in amplitude) or a high AP response as a favorable prognostic criterion for recovery. 1 case of nearly complete sudden deafness with partial recovery reported by *Rietema* [1979] showed at the first examination (14 days after onset) only at 4,000 Hz a minimal –SP at high stimulus intensities. 2 months after being hospitalized, the –SP for the test frequencies of 2, 4, and 8 kHz showed 'low' normal values.

Compound Action Potential

Although most publications report on AP findings, most of them lack valuable information, i. e. electrocochleography obviously was carried out with only one stimulus signal (preferably click stimuli!) or only one intensity [*Graham* et al., 1978; *Graham,* 1979; *Nishida* et al., 1976]. Some authors only published the waveforms of the recorded potentials [*Beagley and Gibson,* 1978]. However, the determination of the amplitude-intensity and latency-intensity curves for compound APs elicited with tone-bursts (or similar stimulus signals, e. g. third-octave clicks or Gaussian-shaped tone pulses) of different frequencies is a sine qua non. A rough separation between retrocochlear and cochlear hearing losses of different origin can be obtained by means of a log amplitude-log latency plot [*Eggermont,* 1976b]: The values of hearing losses of neural origin (e. g. acoustic neuroma) fall well above the 2σ-boundaries of normal hearing, while those due to hair cell loss fall beyond the normal range. Cases of Menière's disease are not distinguishable from normal hearing.

Valuable information as to how a recruitment can reliably be detected is presented solely in several papers from *Eggermont* [1976b, 1977a, b].

The most important findings and their implications have already been described in the preceding sections. Neither the description of the waveform nor I/O curves nor absolute values of amplitude or latency of the compound AP are reliable parameters to decide on the presence of recruitment. The most reliable criterion is the compound AP tuning curve [*Eggermont*, 1977a]: A broadening of the tuning curve along with a normal low-frequency behavior is the electrocochleographic correlate of recruitment. *Eggermont* [1977b] determined tuning curves in several stages of 1 case of sudden hearing loss. The recovery of the hearing loss is directly related to a resharpening of the initially broad tuning curve, until a quite normal shape is reached. The sharp tuning found in cases of deafness of neural origin seems to be preserved (though the threshold at the characteristic frequency is commonly elevated). Tuning curves in cases of cochlear hearing loss due to ototoxic drugs exhibit a similar broad tuning like those in cases of sudden hearing loss. In Menière's ears, some of the tuning curves evenly exhibit the same broad shape, whereas in some cases a certain tip is preserved.

Discussion

Electrocochleographic findings in cases of sudden hearing loss are exceptionally few. This fact may be due to several reasons. One important reason is that – throughout this volume – only the genuine sudden loss of *cochlear* (and vestibular) function is considered. One publication which deals with electrocochleographic findings in 70 patients suffering from sudden sensorineural hearing loss [*Graham* et al., 1978] merely contains one quotable case of obviously genuine sudden hearing loss. This restriction in definition results in another restriction with respect to Electric Response Audiometry: Nothing but (transtympanic) electrocochleography would yield valuable results. Electrocochleography, however, has fallen into disrepute during the past 6 or 7 years. The reasons are obvious: Transtympanic electrode placement is an otological intervention which also involves some degree of risk to the patient. If the audiologist has not got at least some training in ENT (as is the fact in many countries, especially in the United States), he would not be able to employ this technique without medical assistance (also true of local and general anesthesia). The brainstem evoked response audiometry (BERA) which has evinced a marked progress during the past 7 years, however, does not have this disadvantage:

it can be employed not only by physicians of other specialties, but also by nonmedical audiologists. Despite the great importance of BERA, this development is certainly responsible for the considerable lack of electrocochleographic findings. The third important reason is that justified doubts exist about the exposure of an already impaired ear to higher sound intensities.

In this context, some remarks should be made about what the terms 'cochlear' or 'sensory' with respect to hearing disorders refer to. Both terms seem to be used synonymously. However, most physiological findings on cochlear processes have been obtained from single auditory nerve fibre data. The few available data obtained from single (inner) hair cells obviously exhibit quite similar properties and behavior to those of primary auditory neurons (frequency selectivity, toxic and metabolic vulnerability, etc.). It is not our aim to speculate on the localization of this vulnerable 'second filter'. Compound AP data as recorded in man, however, seem to justify the assumption that specified findings obtained from the compound AP point to a sensory origin, especially the (eventually reversible) loss of sharp tuning.

In an electrocochleographic strategy for the diagnosis of idiopathic sudden hearing loss, CM and SP are of inferior value, compared to specified parameters of the compound AP. CM – in the absence of AP – can be used only as a diagnostic hint for idiopathic sudden deafness [*Beagley*, personal communication, 1979; *Graham* et al., 1978; *Naunton and Zerlin*, 1978]. The SP, which is assumed to be the only reliable parameter for estimating hair cell validity in the basal turn of the cochlea [*Rietema*, 1979], may indicate a dysfunction or loss of outer hair cells. An increased –SP magnitude can be useful to exclude cases of Menière's disease when a sensory origin of the hearing loss is assumed. Normal or reduced values, however, do not permit this distinction.

The most important information can be retrieved from the compound AP. A log amplitude-log latency plot allows one to separate hearing losses due to hair cell impairment (data entries fall beyound the 2σ-boundary of normal hearing), while Menière ears cannot be distinguished from normal hearing. This very important finding obviously points to different pathogenetic mechanisms for sudden hearing loss and Menière's disease. The most reliable sign for recruitment, as a typical symptom of genuine sudden hearing loss (if no complete deafness occurs), is a broadening of the compound AP tuning curve along with normal low-frequency behavior. Recovery of hearing is accompanied by a resharpening of the tuning curve.

This finding is very similar to the temporary, reversible detuning of primary auditory neurons [Evans, 1974] or inner hair cells [Sellick and Russell, 1978, 1979] caused by hypoxia, and it supports the hypothesis of a circulatory disturbance as a cause of sudden hearing loss. Hair cell losses, however, due to ototoxic agents or noise exposure, exhibit similar broad tuning curves as does the idiopathic sudden hearing loss. A distinction between these subgroups on the basis of compound AP tuning curves is impossible, and the electrocochleographic diagnosis of sudden hearing loss is, again, a diagnosis per exclusionem.

The determination of compound AP tuning curves is a time-consuming technique which requires extended equipment and, last not least, an experienced investigator. The investigation time could be reduced if the relationship between detuning and reduced narrow-band AP latency could be established.

The derivation of prognostic criteria (dominant –SP, high AP response) from electrocochleographic findings seems, at this stage, to be doubtful. The number of cases as reported by Nishida et al. [1976] is too limited to be more than a support for their hypothesis. 1 case reported by Rietema [1979] showed, even 19 days after onset, hardly a detectable SP, though hearing recovered considerably within 2 months after hospitalization.

The question arises as to whether an electrocochleographic investigation in cases of sudden hearing loss might be useful or not. In the authors' opinion, only very elaborated techniques can give really reliable information. The findings can certainly be very helpful in differential diagnosis unless the degree of hearing loss impedes the collection of necessary data. The most significant value of such data seems to be the scientific one which should not be underestimated. The reliability of an experimental model of sudden hearing loss, which is most important for the understanding of the underlying processes as well as for the evaluation of therapeutic concepts, cannot be developed unless sufficient electrophysiologic findings from man (= electrocochleographic findings) are available.

Conclusion

It is the cochlear dysfunction in particular which highlights the significance and necessity of electrocochleography. An inspection of waveforms and measurements of latencies alone can be misleading and thus give rise to serious misinterpretations. Reliable results can be obtained only by employing more elaborated techniques.

Determining compound AP tuning curves, for example, allows one to distinguish significantly hearing disorders of sensory origin. The question as to whether patients with sudden hearing loss should be subjected to extensive supra-threshold audiometric testing cannot be answered at this stage.

If the departmental policy allows special electrocochleographic investigations, the results may be not only of diagnostic value, but also promote the progress of our scientific cognition of the disease in question.

Summary

Starting from a (heuristic) definition of the genuine sudden hearing loss, this contribution begins with a brief description of the physiological foundations necessary for the understanding of electrocochleographic findings as well as the foundations of the relevant electrocochleographic techniques. Special emphasis is given to the actual neurophysiologic hypothesis of the recruitment phenomenon and the question is raised as to how specified electrocochleographic findings reliably reflect the presence of recruitment. In turn, the outcome of an internationally circulated questionnaire is presented and discussed. The authors underline the importance of electrocochleography, in particular for the diagnosis of cochlear disorders.

References

Allen, J. B.: Cochlear micromechanics – a mechanism for transforming mechanical to neural tuning in the cochlea. J. acoust. Soc. Am. 62: 930–939 (1977).

Aran, J. M.; Charlet de Sauvage, R.: Clinical value of cochlear microphonic recording; in Ruben, Elberling, Salomon, Electrocochleography, pp. 55–56 (University Park Press, Baltimore 1976).

Aran, J. M.; Negrevergne, M.: Aspects cliniques de quelques formes pathologiques particulières des réponses du nerf auditif chez l'homme. Audiology 12: 488–503 (1973).

Aran, J. M.; Pèlerin, J.; Lenoir, J.; Portmann, C.; Darrouzet, J.: Aspects théoriques et pratiques des enregistrements électro-cochléographiques selon la méthode établie à Bordeaux. Revue Laryng., Bordeaux suppl., pp. 601–644 (1971).

Beagley, H. A.: Can we use the cochlear microphonic in electrocochleography? Revue Larygn., Bordeaux 95: 531–536 (1974).

Beagley, H. A.; Gibson, W. P. R.: Electrocochleography in adults; in Naunton, Fernández, Evoked electrical activity in the auditory nervous system, pp. 259–274 Academic Press, New York 1978).

Békésy, G. von: The coarse pattern of the electrical resistance in the cochlea of the guinea pig (electroanatomy of the cochlea). J. acoust. Soc. Am. 23: 18–28 (1951).

Dallos, P.: Cochlear potentials and cochlear mechanics; in Møller, Basic mechanisms in hearing, pp. 335–372 (Academic Press, New York 1973).

Durrant, J. D.; Gans, D.: Biasing of summating potentials. Acta oto-lar. 80: 13–18 (1975).

Eggermont, J. J.: Electrophysiological study of the normal and pathological human cochlea. I. Presynaptic potentials. Revue Laryng., Bordeaux 97: suppl., pp. 487–495 (1976a).

Eggermont, J. J.: Electrocochleography; in Keidel, Neff, Handbook of sensory physiology, vol. V, part 3, pp. 625–705 (Springer, Berlin 1976b).

Eggermont, J. J.: Summating potentials in electrocochleography: relation to hearing disorders; in Ruben, Elberling, Salomon, Electrocochleography, pp. 67–87 (University Park Press, Baltimore 1976c).

Eggermont, J. J.: Analysis of compound action potential responses to tone bursts in the human and guinea pig cochlea. J. acoust. Soc. Am. 60: 1132–1139 (1976d).

Eggermont, J. J.: Electrocochleography and recruitment. Ann. Otol. Rhinol. Lar. 86: 138–149 (1977a).

Eggermont, J. J.: Compound action potential tuning curves in normal and pathological human ears. J. acoust. Soc. Am. 62: 1247–1251 (1977b).

Eggermont, J. J.: Compound action potentials: Tuning curves and delay times; in Hoke, de Boer, Models of the auditory system and related signal processing techniques. Scand. Audiol. suppl. 9: pp. 129–139 (1979a).

Eggermont, J. J.: Summating potential in Menière's disease. Archs Oto-Rhino-Lar. 222: 63–75 (1979b).

Eggermont, J. J.; Odenthal, D. W. Frequency selective masking in electrocochleography. Revue Laryng., Bordeaux 95: 489–496 (1974).

Eggermont, J. J.; Odenthal, D. W.: Frequency selective masking in electrocochleo-basic principles and clinical application. Acta oto-lar. suppl. 316, pp. 1–84 (1974).

Elberling, C.: Action potentials along the cochlear partition recorded from the ear canal in man. Scand. Audiol. 3: 13–19 (1974).

Evans, E. F.: Does frequency sharpening occur in the cochlea? Symp. on Hearing Theory, IPO Eindhoven 1972, pp. 27–34.

Evans, E. F.: The effects of hypoxia on the tuning of single fibres in the cochlear nerve. J. Physiol., Lond. 238: 65–67 (1974).

Evans, E. F.: The sharpening of cochlear frequency selectivity in the normal and abnormal cochlea. Audiology 14: 419–442 (1975).

Feldmann, H.: Sudden hearing loss, a clinical survey (this volume).

Fowler, E. P.: Marked deafened areas in normal ears. Archs Otolar. 8: 151–155 (1928).

Gibson, W. P. R.: Clinical electrocochleography: the significance of the summating potential in Menière's disorder; in Barber, Evoked potentials (MTP Press, Lancaster 1980).

Goldstein, J. L.: Auditory nonlinearity. J. acoust. Soc. Am. 41: 676–689 (1967).

Graham, J. M.: Tinnitus and deafness of sudden onset. Electrocochleographic findings in 100 patients. Paper presented 1st Int. Tinnitus Seminar, New York 1979.

Graham, J. M.; Ramsden, R. T.; Moffatt, D. A.; Gibson, W. P. R.: Sudden sensori-neural hearing loss: electrocochleographic findings in 70 patients. J. Lar. Otol. *92:* 581–589 (1978).

Handrock, M.: Die Reaktionsform der Kochlea beim experimentell induzierten Hör-sturz. Lar. Rhinol. *57:* 881–891 (1978).

Hieke, D.; Hoke, M.: Influences of the transfer function of the middle ear on the vectorial integral of cochlear microphonics (compound CM); in Hoke, von Bally, Proc. Conf. Electrocochleography and Holography in Medicine, pp. 105–114 (Sonderforschungsbereich 88, Münster 1976).

Hoke, M.: Cochlear microphonics in man and its probable importance in objective audiometry; in Ruben, Elberling, Salomon, Electrocochleography, pp. 41–54 (University Park Press, Baltimore 1976a).

Hoke, M.: Problem of cochlear microphonic recording in man. Revue Laryng. Bor-deaux *97:* suppl., pp. 473–486 (1976b).

Hoke, M.; Hieke, D.: The significance of cochlear microphonics recorded with a gross electrode ('compound' CM); in Hoke, von Bally, Proc. Conf. Electro-cochleography and Holography in Medicine, pp. 95–104 (Sonderforschungs-bereich 88, Münster 1976).

Honrubia, V.; Strelioff, D.; Sitko, S.: Electroanatomy of the cochlea: its role in cochlear potential measurements; in Ruben, Elberling, Salomon, Electrocochleo-graphy, pp. 23–29 (University Park Press, Baltimore 1976).

Jung, W.; Rosskopf, K.: Evoked Response Audiometry (ERA) am Meerschweinchen vor und nach Lasix-induziertem Hörsturz. Lar. Rhinol. *54:* 411–418 (1975).

Kiang, N. Y.-S.; Moxon, E. C.; Levine, R. A.: Auditory-nerve activity in cats with normal and abnormal cochleas; in Wolstenholme, Knight, Sensorineural hearing loss, pp. 241–273 (Churchill, London 1970).

Kohllöffel, L. U. F.: Studies on the distribution of cochlear potentials along the basilar membrane. Acta oto-lar. suppl. 288, pp. 1–66 (1971).

Köpcke, J.; Hoke, M.; Lütkenhöner, B.: Influence of the intratympanic record-ing site on the frequency response of cochlear microphonics; in Hoke, Kauff-mann, Bappert, Cochlear and brainstem evoked response audiometry and electrical stimulation of the VIIIth nerve. Scand. Audiol. suppl. *11:* pp. 65–71 (1980).

Kupperman, R.: The SP in connection with movements of the basilar membrane; in Plomp, Smoorenburg, Frequency analysis and periodicity pitch detection in hearing, pp. 126–131 (Sijthoff, Leiden 1970).

Lindsay, J. R.: Histopathology of Menière's disease as observed by light microscopy. Otolaryngol. Clin. North Am. *1:* 319–329 (1968).

Moffat, D. A.: Transtympanic electrocochleography in Menière's disease: Variation in the amplitude of the summating potential related to clinical status. Brit. J. Audiol. *13:* 149–152 (1979).

Moffat, D. A.; Gibson, W. P. R.; Ramsden, R. T.; Morrison, A. W.; Booth, J. B.: Transtympanic electrocochleography during glycerol dehydration. Acta oto-lar. *85:* 158–166 (1978).

Morrison, A. W.; Moffat, D. A.; OConnor, A. F.: Clinical usefulness of electro-cochleography in Menière's disease: an analysis of dehydrating agents. Oto-laryngol. Clin. North Am. *13:* 703–721 (1980).

Naunton, R. F.; Zerlin, S.: Basis and some diagnostic implications of electrocochleography. Laryngoscope 86: 475–482 (1976).

Nishida, H.; Kumagami, H.; Dohi, K.: Prognostic criteria of sudden deafness as deduced by electrocochleography. Archs Otolar. 102: 601–607 (1976).

Rietema, S. J.: The clinical significance of electrocochleography; thesis, Leiden
(1979).

Ruben, R. J.: Cochlear potentials as a diagnostic test in deafness; in Graham, Sensorineural hearing processes and disorders, pp. 313–337 (Little, Brown, Boston
1967).

Schuknecht, H. F.: Pathology of Menière's disease. Otolaryngol. Clin. North Am. 1:
331–337 (1968).

Schuknecht, H. F.: Pathology of the ear (Harvard University Press, Cambridge 1974).
1974).

Sellick, P. M.; Russell, I. J.: Intracellular studies of cochlear hair cells: filling the
gap between basilar membrane mechanics and neural excitation; in Naunton,
Fernández, Evoked electrical activity in the auditory nervous system, pp. 113–
137 (Academic Press, New York 1978).

Sellick, P. M.; Russell, I. J.: Two-tone suppression in cochlear hair cells. Hearing
Res. 1: 227–236 (1979).

Teas, D. C.; Eldredge, D. H.; Davis, H.: Cochlear responses to acoustic transients:
an interpretation of whole nerve action potentials. J. acoust. Soc. Am. 34: 1483–
1489 (1962)

Whitfield, I. C.; Ross, H. F.: Cochlear microphonics and summating potentials and
the outputs of individual hair cell generators. J. acoust. Soc. Am. 38: 126–131
(1965).

Yoshie, N.: Auditory nerve action potential responses to clicks in man. Laryngoscope, St Louis 78: 198–215 (1968).

Zerlin, S.; Naunton, R. S.: CM in clinical diagnosis; in Naunton, Fernández, Evoked
electrical activity in the auditory nervous system, pp. 275–284 (Academic Press,
New York 1978).

Prof. Dr. M. Hoke, Experimental Audiology, Ear, Nose and Throat Clinic,
University of Münster, Kardinal-von-Galen-Ring 10, D-4400 Münster (FRG)

Adv. Oto-Rhino-Laryng., vol. 27, pp. 100–109 (Karger, Basel 1981)

Difference Limen for Intensity in Patients with Sudden Deafness and Other Inner Ear Disorders[1]

K. Schorn

ENT Clinic of the University, München, FRG

Introduction

The detection of frequency differences, time differences, fine time structures such as roughness and differences in rhythmic structures, for example, are very important in discriminating language and music. Another fundamental property of hearing is the ability of the ear to discriminate intensity differences. The most widely used clinical test for evaluating the capacity of the ear to discriminate intensity differences is most certainly the SISI test (short increment sensitivity index) which is characterized in that a pure tone is presented to the patient at a specified sensation level (SL) and a small increase in intensity is superimposed upon the steady-state tone at periodic intervals [1]. Various psychoacoustic experiments have indicated, however, that it would be more expedient to measure the patient's precise difference limen for intensity which is not explored in the SISI test [2, 3]. In the procedure described in the following, two tone pulses are presented which differ in intensity and which are separated by pauses.

Method

The apparatus for measuring the difference limens for intensity (DLI) with pulsed tones which was used throughout this investigation was de-

---

[1] Dedicated to Prof. Dr. med. *A. Herrmann* on his 80th birthday.

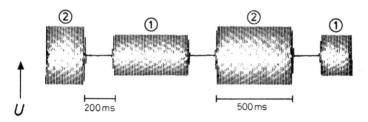

Fig. 1. Time domain waveform of the stimulus signal showing the experimental paradigm of alternately presented pure tone pulses [from ref. 5].

veloped by *Fastl and Zwicker* [4] and, with our collaboration, was simplified for clinical use. Since a detailed description of the equipment has been published elsewhere [4], this paper will be restricted to the essentials. Pulsed tones with a duration of 500 ms and a Gaussian-shaped rise/decay time of 20 ms are separated by pauses of 200-ms duration (fig. 1). In consideration of possible cases involving hearing deterioration in a restricted frequency region, the following six frequencies were selected for the investigations: 250, 500, 1,000, 2,000, 4,000 and 8,000 Hz. The levels of the alternating test tones can be increased in increments of 0.5 dB.

The pure tone threshold for the test frequencies was determined with an accuracy of 5 dB, and the determination of the difference limen for intensity was performed at a sensation level of 30 dB. The patient was merely requested to indicate whether the loudness of the successive tone pulses was the same or different. Like the SISI test, the patients require a conditioning phase. Tone pulses of equal intensity on the one hand and, on the other hand, tone pulses featuring large intensity differences, e. g. 10 dB, are presented. Only then can the experimenter gradually approach the critical value at which the intensity difference becomes just noticeable.

Results

Normative values of the DLI were initially determined in normal hearing subjects who had been hospitalized or were undergoing outpatient treatment for other complaints. To ensure that the patients in fact had normal hearing, speech audiometry and tympanometry were performed to supplement the pure tone audiometry. The medians for the difference

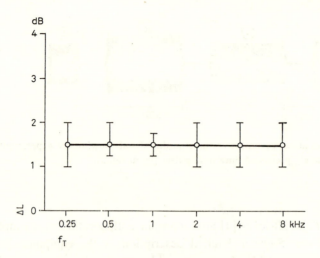

Fig. 2. Just noticeable differences in intensity of pulsed tones, expressed as $\varDelta L$, as a function of frequency. Normative values, obtained from 97 normal hearing listeners [from ref. 5].

limen for intensity amounted to 1.5 dB at all frequencies, thus indicating that they are independent of frequency (fig. 2) [5]. This median was determined in subjects in all age groups, various professions and social status. The average age was 33 years. A slightly improved value of 1.25 dB was found in young, trained listeners. This value was within the standard deviation of our normal hearing subjects. Musical aptitude played no role whatsoever.

Difference limens for intensity in patients suffering from cochlear impairments were examined in the following audiological disorders: sudden deafness, noise-induced hearing loss, childhood hearing impairment, hearing impairments caused by ototoxic agents, Menière's disease and presbyacusis [5].

In the 13 patients suffering from *sudden deafness* – average age 43 years – the difference limens for intensity were slightly elevated at all frequencies (fig. 3). In the frequency range up to 1,000 Hz, the medians of the difference limens for intensity varied between 2 and 2.5 dB and were somewhat lower at higher frequencies. Regarding the individual data from patients in this group, DLI can be found comparable to those of normal hearing subjects.

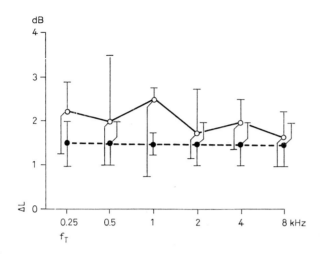

Fig. 3. Just noticeable differences in intensity in 13 patients suffering from sudden deafness (open circles). The normative values are drawn for comparison (solid circles) [from ref. 5].

The DLI were clearly independent of the hearing loss which amounted to an average of 55 dB HL at 500 Hz and 52.5 dB HL at 4,000 Hz. The same data as in cases with a slight high-frequency loss were also found in patients suffering from moderate to profound deafness as well as in cases showing a pancochlear hearing loss of about 60 dB. The SISI test was positive in all examined patients with sudden deafness, i. e. intensity differences of 1 dB were detected.

The DLI in patients with cochlear hearing loss of different origins did not show any substantial difference compared to sudden deafness. The medians of the DLI of 20 patients with chronic *noise-induced hearing loss* – average age 47 years – were slightly elevated throughout the whole frequency range, although to a lesser extent than in the patients suffering from sudden deafness (fig. 4). Individual findings are comparable in this case as well to those in normal hearing subjects. The salient feature is that there is no difference in DLI between the lower frequency range, showing an average hearing loss of 15 dB at 500 Hz and the frequency of 4000 Hz, with 45 dB exhibiting the greatest hearing loss on the average.

Childhood hearing impairment and hearing impairment caused by ototoxic agents – 15 cases with an average age of 37 years – revealed

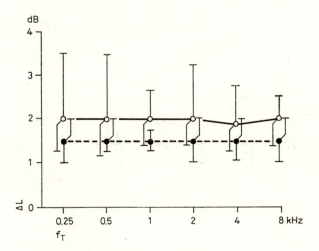

Fig. 4. Just noticeable differences in intensity in 20 patients suffering from noise-induced hearing loss (open circles). The normative values are drawn for comparison (solid circles) [from ref. 5].

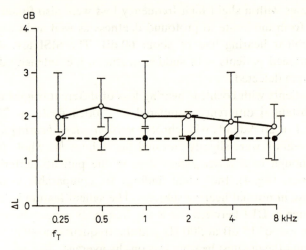

Fig. 5. Just noticeable differences in intensity in 15 patients suffering from oto-toxic hearing impairment (open circles). The normative values are drawn for comparison (solid circles) [from ref. 5].

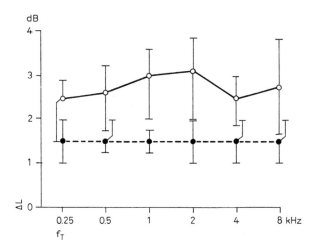

Fig. 6. Just noticeable differences in intensity in 21 patients suffering from Menière's disease (open circles). The normative values are drawn for comparison (solid circles) [from ref. 5].

slightly elevated medians of the DLI varying from 1.75 to 2.2 dB (fig. 5). Compared to the normative values, the DLI was scarcely elevated at 8 kHz; the median at this frequency, however, comprised only nine individual data, since the rest of the patients had suffered from such a severe hearing loss that examination of the DLI at 8 kHz was no longer possible. The group of test subjects included patients with deafness acquired after virus infections such as measles, mumps, etc., or after treatment with ototoxic medication, such as gentamycin.

We found somewhat higher DLI in the 21 patients – average age 46 years – suffering from *Menière's disease* which had increased to 2.5–3 dB (fig. 6). The mean hearing loss amounted to 55 dB at 500 Hz and to 45 dB at 4,000 Hz. We also found in this group of patients that the increase in the DLI is irrespective of the hearing loss.

Slightly elevated DLI were again found in the group of patients with *presbyacusis*. In the 34 patients with an average age of 66 years, the median of the difference limen for intensity was 2.5 dB at most frequencies and somewhat less only at 4,000 Hz (fig. 7). The hearing loss, on the other hand, was especially pronounced at high frequencies. It amounted on the average to 57.5 dB at 4,000 Hz, but only 30 dB at 500 Hz.

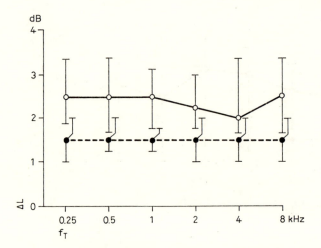

Fig. 7. Just noticeable differences in intensity in 34 patients suffering from presbyacusis (open circles). The normative values are drawn for comparison (solid circles) [from ref. 5].

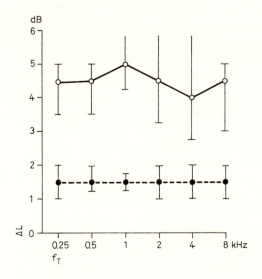

Fig. 8. Just noticeable differences in intensity in 15 patients suffering from retro-cochlear hearing impairment (open circles). The normative values are drawn for comparison (solid circles [from ref. 5].

Table 1. Comparison of just noticeable differences in intensity in normal hearing listeners and in different groups of subjects suffering from hearing impairments of different origin

Group	DLI, dB	Number
Normal hearing	1.50	97
Sudden deafness	2.02	13
Noise-induced hearing loss	1.98	20
Ototoxic hearing impairment	1.97	15
Menière's disease	2.75	21
Presbyacusis	2.38	34
Retrocochlear hearing impairment	4.50	15

A marked increase in the difference limens for intensity could be found in all patients with *retrocochlear hearing impairment* (fig. 8). The median values varied between 4.5 and 5 dB, i. e. 3.5 dB higher than the normative values and an average of 2 dB higher than the values in cochlear impairments. The 15 patients with retrocochlear impairment included 10 patients suffering from an acoustic neuroma, 2 patients with brain injury as a result of an accident, 2 patients suffering from apoplectic stroke and 1 patient with cerebral haemorrhage. The average age was 43 years; the pure tone threshold was elevated to 42.5 dB HL at 500 Hz and to 70 dB HL at 4,000 Hz. The SISI test revealed recruitment in 8 patients and failed in 6 patients. 2 patients did not undergo testing as their hearing loss was very minor, i. e. less than 40 dB HL.

Discussion

The DLI can be determined in two entirely different ways: with amplitude-modulated tones as in the SISI test or Lüscher test, for example, or with alternately pulsed tones of different intensity [6]. The most widely used suprathreshold test is certainly the SISI test in which the detection of slight amplitude modulations is considered an achievement of the inner hair cells, whereas the outer hair cells do not perceive intensity differences [7]. Doubts have recently been voiced, however, about the significance of the SISI test in differentiating cochlear from retrocochlear damage [8–10].

A test was therefore developed in which the patient must discriminate tone pulses of different intensity with a relatively long interposed pause

(200 ms) on the assumption that in this procedure the loudness is not compared in the auditory periphery, but rather in more central parts of the auditory system which are probably responsible for discriminating the loudness. Based on these assumptions, we expected poorer detection of the intensity differences of the pulsed tones in patients with retrocochlear impairment than with cochlear deafness. As the results show (table I), this is in fact the case. Patients with retrocochlear impairment caused by an acoustic neuroma, apoplectic stroke, brain injury or cerebral haemorrhage all show a clearly elevated DLI compared not only to normal hearing subjects but also to the DLI in patients with auditory dysfunctions of different cochlear origins. This SISI test, by contrast, is negative in only 6 of the 15 patients suffering from retrocochlear impairments, i. e. 37.5 %. In 8 patients, in particular those suffering from an acoustic neuroma, recruitment determined by means of intensity-modulated test tones was positive. The test was not able to be performed on 2 patients as the hearing loss was too insignificant.

In patients who had suffered from sudden deafness, noise-induced hearing loss and hearing impairments caused by ototoxic agents, there is a slight elevation of the DLI compared to normal hearing subjects, although this is not significant. This elevation in the DLI determined with pulsed tones in patients with cochlear deafness, although only slight, nonetheless shows that this test cannot be compared directly with the SISI test. The SISI test was positive in all patients with cochlear impairment, i. e. low DLI amounting to only 1 dB were detected.

We found an elevation of the DLI using pulsed tones in patients suffering from presbyacusis and Menière's disease. There is a simple explanation as far as presbyacusis is concerned: since the cerebral blood supply is frequently affected in patients with cerebral sclerosis, deafness thus comprises peripheral and central components. The elevated DLI could not be explained in patients with Menière's disease. We initially thought that the complaints of vertigo affected the accuracy of the examination. A check of the individual data, however, revealed that elevated DLI were also found in these patients during a period in which they suffered no vertigo. In order to clarify the situation, measurements should again be performed on this group of patients.

It can be said in summary that the measurement of DLI with an interposed pause is a psychoacoustic test which makes it possible to differentiate significantly between cochlear and retrocochlear damage. It should thus be incorporated into the battery of suprathreshold audiometric tests.

We see a distinct advantage over the SISI test in that the examinations can also be carried out on patients with a minor hearing loss less than 40 dB.

Summary

A technique was developed for determining the difference limens for intensity (DLI) with pulsed tones. With this method, DLI in patients with sudden deafness, Menière's disease, noise-induced hearing loss, presbyacusis, ototoxic impairment as well as retrocochlear impairment is determined and compared with the data from normal hearing subjects.

The advantages of this method over the SISI test which uses amplitude-modulated tones are: (a) retrocochlear disorders can be detected with great reliability, and (b) the test can also be used in patients with normal hearing or a hearing loss of less than 40 dB.

References

1   Jerger, J. F.; Shedd, J. L.; Harford, E.: On the detection of extremely small changes in sound intensity. Archs Otolar. 69: 200–211 (1959).

2   Zwicker, E.: Die Grenzen der Hörbarkeit der Amplitudenmodulation und der Frequenzmodulation eines Tones. Acustica, Akustische Beihefte AB 125–AB 133 (1952).

3   Zwicker, E.: Scaling; in Keidel, Neff, Handbook of sensory physiology, vol. V/2, pp. 401–448 (Springer, Heidelberg 1975).

4   Fastl, H.; Zwicker, E.: A device for measuring level and frequency difference limes. J. Audiol. Technique 18: 26–34 (1979).

5   Fastl, H.; Schorn, K.: Discrimination of level differences by hearing impaired patients. Audiology (in press).

6   Denes, P.; Naunton, R. F.: The clinical detection of auditory recruitment. J. Laryngol. 64: 375–398 (1950).

7   Lehnhardt, E.: Praktische Audiometrie; 5. Aufl. (Thieme, Stuttgart 1978).

8   Eggermont, J. J.: Electrocochleography and recruitment. Ann. Otol. 86: 138–149 (1977).

9   Lamoré, P. J.; Rodenburg, M.: Significance of the SISI test and its relation to recruitment. Audiology 19: 7585 (1980).

10 Owens, E.: Audiologic evaluation in cochlear versus retrocochlear lesions. Acta oto-lar. suppl. 283 (1971).

Prof. Dr. Karin Schorn, ENT Clinic of the University München,
Marchioninistrasse 15, D-8000 München 70 (FRG)

Adv. Oto-Rhino-Laryng., vol. 27, pp. 110–113 (Karger, Basel 1981)

Personality System and Sudden Deafness: A Comparative Psychological Study

Jochen Dohse, Siegfried Lehrl, Michael Berg

HNO-Universitätsklinik Erlangen, Erlangen, FRG

Introduction

Persons suffering from 'sudden deafness' are often said to be conspicuous by their psyche or it is found that they are in a psychic crisis at the time of the illness. These are the catch-words that have led to this study. They reflect the impression of the attending doctor and are based on his own professional background. The aim of this work is to find out whether there is a correlation between this personal impression and the personality system of the patient and, secondly, whether this personality system actually resembles the psychic structure of psychosomatic patients.

Methods

There exist various psychological tests which differ according to their respective tasks: in psychiatry, for example, to diagnose delusion or in marketing research to examine the efficiency of advertising. They also differ insofar as some of them are more oriented towards the momentary state of health and feeling whereas others measure the general, more or less time-independent structure of the personality. For this study the 'Freiburger Persönlichkeitsinventar' (FPI) has been used, a test supposed to evaluate a personality regardless of the respective present state of health.

This test is particulary appropriate, because we do not want to make a statement about the patient's condition after becoming ill. The test should measure the patient's personality before having fallen ill. Several works have proved the stability of this test. It is mostly used in psycho-

therapy and in psychosomatics consisting of 114 statements, e. g. 'I think
I could be a passionate hunter' or 'sometimes I think I'm good for nothing' or 'in company I usually behave better than at home'. The patient
expresses his opinion to each of these statements by ticking true or false.
The analysis shows certain components in twelve different divisions; such
as 'sociability', 'depression', 'neuroticism', 'introversion' or 'extraversion'.
The FPI is known not to be influenced if, for example, patients are put
into hospital, where they are exposed to an unfamiliar environment. It
could happen, however, that certain striking features develop as soon as a
patient becomes hard of hearing on one ear. For this reason a control
group of patients with defective hearing on one ear due to otosclerosis
was also examined.

In addition to the FPI an intelligence test, the 'Mehrfachwahl-Wort-
schatz-Intelligenztest' (MWT-B), was presented to the patients. It gives
information about the general level of intelligence. In 37 lines consisting
of five words one word was taken from the colloquial-, culture- or science
language. The other four words are senseless, e. g. 'siziol-salzahl-sozihl-
sziam-sozial'. The patient's job is to mark the word which he believes to
know.

In the course of 1 year we examined 35 patients suffering from otosclerosis as well as 30 patients suffering from sudden deafness; 17 patients
of the second group had idiopathic sudden deafness while the other 13
had other diseases which in principle could be made responsible for sudden deafness.

Results

The patient's intelligence measured by the MWT-B showed no remarkable deviation from normal and no differences between the two
groups of patients either. In the FPI test the patients suffering from otosclerosis proved to be inconspicuous. This leads to the conclusion that
one-sided deafness does not cause psychic changes; a possibility that could
elsewise cause a systematic error, falsifying the results of this study.

Among the 13 patients suffering from other diseases contributing to
sudden deafness 7 had become deaf in the course of an infection. They
did not show any psychic changes either. 6 patients with sudden deafness
were suffering from additional organic, cardiac or circulatory disturbances.
They showed deviations from the average population in the sense of

emotional disorder. Their results were in full accordance with those quoted by the respective test authors for patients with similar organic cardiac and circulatory diseases. For example, after a cardiac infarct the patient's living conditions change and this fact results in a change of his personality.

Patients with idiopathic sudden deafness had a tendency to deviate in the FPI from the average population; they had a tendency to lack self-confidence and stability of mood. In the case of psychic agitation they suffered from vegetative disorders such as sweating attacks in a proportion that is above average. But if we take into consideration the success of therapy we find striking differences in the results: patients with improved hearing described themselves as 'self-possessed', 'sociable', 'enterprising' and 'making their way'. Their average mood was balanced with only a slight tendency to increased sorrows and guilt complexes. The point is stressed here again: This reflects general time-independent personality profile – not the current state of health and feeling.

On the other hand, patients without improved hearing after treatment, or even with deteriorated hearing, showed results significantly above average in four divisions: 'nervousness', 'depression', 'inhibition' and 'neuroticism' and results significantly below average in the division 'masculinity'. In other words these patients are people who suffer from intensive vegetative symptoms in the state of emotional excitation; they are often depressed or have deep guilt or inferiority complexes. They are inhibited when getting into contact with other people, they have difficulty in asserting themselves, are susceptible and often feel that they are being treated unjustly. They also have little self-reliance and confidence. This personality system is in full accordance with the one that the test authors describe for psychosomatic patients.

Conclusion

Deafness on one ear does not cause any significant change of personality as could be shown on the basis of the results of the otosclerosis patients. Patients with sudden deafness suffering from additional organic diseases do not show any psychic changes (like patients with infectious diseases); if there are any psychic changes, however, they are due to a basic disease such as a cardiac or circulatory defect.

Patients suffering from idiopathic sudden deafness can be differentiated into two groups: one group consists of persons with no psychic disease. In this case organic therapy as well as the tranquillity in the hospital leads to recovery. Talks with these patients made clear that temporary private or professional strain can be made responsible for this disease. The other group consists of persons with the above-mentioned neurotic personality system. Here organic therapy and tranquillity in hospital are obviously not enough to restore hearing. Here one could try to help them to cope with their conflicts in psychotherapeutical talks. Thus, a certain relaxation could perhaps contribute to the clinical therapy to restore the patient's hearing.

Summary

The psychic structure of 30 patients suffering from sudden deafness and of 35 patients suffering from otosclerosis was tested by the FPI. Additionally the intelligence was tested by the MWT-B. It could be shown that deafness on one ear does not cause any significant change of personality. Patients with sudden deafness and additional organic diseases did not show any psychic changes or these changes are due to a basic disease. Patients suffering from idiopathic sudden deafness can be differentiated into two groups: the first group consists of patients with no psychic disease; these patients show a good recovering tendency from sudden deafness. The second group consists of patients with a slightly neurotic personality which is in full accordance with those of psychosomatic patients. In these case the organic therapy did not lead to recovery.

J. Dohse, MD, HNO-Universitätsklinik Erlangen, Waldstrasse 1,
D-8520 Erlangen (FRG)

Adv. Oto-Rhino-Laryng., vol. 27, pp. 114–120 (Karger, Basel 1981)

Acute Acoustic Trauma

B. Kellerhals

Universitäts-HNO-Klinik, Inselspital Bern, Switzerland

In sharp contrast to sudden deafness, acoustic trauma can be studied experimentally. Thus, in the literature concerning acoustically induced inner ear damage, an overwhelming amount of experimental and clinical data has been collected. In the present context it is quite impossible to cover all aspects of these damages, presenting a complete review of our – still limited – knowledge. The aim of this paper has to remain a limited one: it should give the reader a summary of the main facts and of the unsolved problems which are of prime *clinical* interest. Such a clinical interest concentrates necessarily on problems of pathogenesis and on the crucial question whether acute acoustic damage is suited for attempts at therapy, or whether it has to be considered a hopeless situation for any efforts at therapy.

Terminology

According to the point of view, acute acoustic inner ear damages can be grouped differently; a patient-oriented approach necessarily leads to another classification than an approach that stresses the scientific fundamentals. The classical terminology of *Rüedi and Furrer* [1946] needs revision. The differentiation between report trauma and explosion trauma, as it was fixed at 1.5 ms duration of the impulse peak, has lost its importance, and chronically induced noise damage should not be termed trauma. *Kellerhals* [1972] and most of the Anglo-American literature differentiate between acute acoustic trauma and chronic noise-induced hearing loss, the latter being synonymous with professional deafness. Such a simplified differentiation will do for practical clinical purposes, but it lacks precision as regards the pathogenetic mechanisms involved. The term 'acute acoustic trauma' includes quite heterogenous pathogenetic mechanisms such as damage following impulse events (defined as acoustic events of less than 1 s duration), short-time exposure to excessive noise intensities, blunt head trauma, and even cases of acoustically triggered sudden hearing losses. Therefore, *Spoendlin* [1980] proposed a classification which differentiates between specific and non-specific inner ear damage, specific damage being localized according to the damaging noise frequencies, whereas non-specific damage shows a pancochlear spread. Such pancochlear damage after minor noise exposure seems to be caused by inner ear microcirculation impairment rather than by the mechanical energy of the triggering acoustic event. After the original description of *Boenninghaus* [1959], such non-specific noise damage was described as *acoustic accident* [*Becker and Matzker*, 1961], *micronoise trauma* [*Gravendeel and Plomp*, 1960], *industrial sudden deafness* [*Kawata*, 1967], *akute Lärmschwerhörigkeit* [*Lehnhardt*, 1965], *akuter Lärmschaden* [*Birnmeyer*, 1970]. Such acoustic accidents – which preferably should be called *acoustically triggered sudden deafness* – are frequently seen. They do not always show a classical flat threshold curve, and there are many transitional stages between clear acoustic trauma and clear sudden deafness.

Pathogenesis of Inner Ear Damage in Acute Acoustic Trauma

Experimental morphology shows that acute acoustic overstimulation is followed by two different types of inner ear alterations. The first group can easily be explained by mechanical forces producing *micro-mechanical lesions* (vs macromechanical lesions, e. g. of the ear drum by explosions) of the organ of Corti such as ruptures of sensory cell membranes, distortion of outer hair cells, separation of supporting cells and of the tympanic layer from the basilar membrane, cilia alterations, rupture of Reissner's membrane. The second group of noise-induced inner ear alterations cannot be caused by direct mechanical energy: these changes include proliferation and vacuolization of the endoplasmic reticulum, swelling of hair cell nuclei, swelling of efferent dendrites below the inner hair cells.

Such a fundamental differentiation between two completely different types of morphological lesions is now generally accepted. Lesions that cannot be explained by mechanical forces suggest some kind of metabolic processes. They are considered the morphological correlate of *metabolic decompensation*. The organ of Corti cannot transform mechanical acoustic energy into the electrical energy of nerve fiber spikes without considerable energy consumption, whereas a microphone performs the same energy transformation without additional energy input. Therefore, acoustic overstimulation necessarily could lead to decompensation of the metabolism of the organ of Corti. This hypothesis is supported substantially by the fact that experimental hypoxia [*Spoendlin,* 1969] produces almost identical morphological changes.

The functional implications of these morphological changes cannot be fixed yet in detail. Of obvious clinical interest is the question of which changes have to be considered as being reversible, and how far such alterations could be influenced by therapeutical means. Mechanical lesions such as a ruptured outer hair cell obviously mean an irreversible loss, although recent experimental work on mechanical lesions of the organ of Corti has shown an unexpected healing potential of some inner ear structures. On the other hand, most of the metabolic changes have to be regarded as reversible alterations.

These considerations suggest that mechanical lesions prevail in permanent hearing loss, whereas reversible metabolic changes easily could explain the temporary threshold shift. Unfortunately, this hypothesis cannot be upheld as a general rule: our present knowledge about the spontaneous evolution of the hearing loss and of the morphologically detectable damage after acute acoustic trauma (fig. 1) contradicts such an assumption. Morphological lesions – especially clear hair cell losses – seem to increase considerably over several weeks, and retrograde degeneration of the peripheral neurons takes as much time as 2–3 months. Metabolic changes tend to recover in a much shorter time, and the temporary threshold shift component of the hearing loss recovers even faster. Figure 1 does not take into consideration that the ratio between mechanical and metabolic lesions depends on the noise intensity: *Spoendlin and Brun* [1973] showed in the guinea pig that mechanical lesions prevail after noise exposure with intensities exceeding 130 dB, whereas lower noise intensities lead mainly to metabolic changes. *Spoendlin* [1980] stressed the point that metabolic changes occur after a certain latency only, and that they are much more variable than mechanical lesions. In acute acoustic trauma, however, impulse noise prevails, and in impulse noise exposures the physical parameters of the acoustic event and its clinical consequences correlate much less than after exposures to steady noise. Impulse noise is definitely

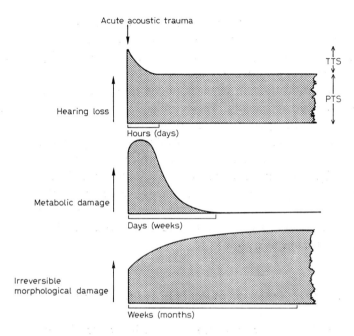

Fig. 1. Time course of the spontaneous evolution of functional damage (hearing loss), reversible metabolic damage and irreversible morphological damage after an acute acoustic trauma. TTS = Temporary threshold shift; PTS = permanent threshold shift. The recovery of metabolic damages takes more time than the spontaneous disappearance of (normal and pathological) TTS. The permanent hearing loss has stabilized before the irreversible morphological damage has reached its final extent.

more dangerous to the inner ear than steady noise of comparable acoustic energy. The protective middle ear reflexes (their effect probably does not exceed 15 dB) do not explain this fact sufficiently. In spite of all these data, the real morphological or metabolic correlate of temporary threshold shift still escapes our knowledge. The temporary threshold shift probably does not rely on mechanical lesions but on some still unknown reversible biochemical processes. As shown in figure 1, the time course of the metabolic changes is not congruent with the time course of temporary threshold shift which is present immediately after the trauma, and which normally subsides within 16 h by exponential decrease. Pathological temporary threshold shifts are often seen in cases of impulse noise damage. They recover more slowly and show a linear regression course. The differentiation between normal and pathological temporary threshold shift

has been used for screening tests in order to detect individuals with abnormal sensitivity to acoustic damage. No such test, however, has proved
sufficiently valuable for singling out persons which are especially prone to
acute acoustic trauma by impulse events.

In acute acoustic trauma, changes of the inner ear microcirculation
have been described repeatedly. These findings, however, are contradic-
tory in many respects, especially in the localization of the microcirculatory
impairment. Microcirculation disturbances thus seem to play a minor role
in acute acoustic trauma, except for cases which have to be considered
acoustically triggered sudden deafness cases. A further problem concerns
the possibility of very late improvement many weeks or months after an
acute acoustic trauma. Such late recoveries suggest an unexpected healing
potential of the organ of Corti – at least in some cases. The base of such
recoveries still remains unknown.

Fundamentals of Treatment in Acute Acoustic Trauma

As can be deducted from figure 1, some inner ear elements pass a
potentially reversible stage before they perish irreversibly. These elements
– which probably are struck with metabolic damage only – theoretically
could be saved by appropriate therapeutical measures. The therapeutical
range lies between spontaneously reversible changes and irreversible mechanical damage. The chance for success is best immediately after the
trauma and decreases slowly during the following days and weeks.

The treatment aims at increasing the inner ear blood flow, thus improving the metabolism of damaged inner ear elements. Many means have
been proposed for this purpose (table I). Among the vasodilating agents,
oxycarbon inhalations (95 % O_2 and 5 % CO_2) seem to be the most effective means for increasing the oxygen supply toward the inner ear [Fisch
et al., 1976], whereas vasoactive drugs such as papaverin and nicotinic
acid cannot be recommended because of their prevailing peripheral effect
which produces no useable increase of the intracranial blood flow [Herr-
schaft, 1975].

Among the microcirculation improving agents, low molecular weight
dextran (1,000 ml the first day, 500 ml the following days) seems to be
the most promising means. Its therapeutic effect has been proved clinically
by Kellerhals [1972], Jakobs and Martin [1977] and Martin and Jakobs
[1977]. Pentoxifyllin seems to increase the inner ear blood flow, too [Kel-

Table I. List of means which were proposed for treatment of acute acoustic trauma

Vasodilatation	nicotinic acid derivates
	xanthin derivates
	papaverin and derivates
	adenosin triphosphate
	blockades of the stellate ganglion
	CO_2 inhalations
	$NaHCO_2$ perfusions
Improvement of microcirculation	low molecular weight dextran
	bencyclan
	pentoxifyllin
Increase of O_2 saturation	hyperbaric oxygen therapy
Various drugs	corticosteroids
	heparin
	procain
	secale alcaloids
	vitamines

lerhals, 1978]. The tinnitus as side effect of acoustically induced hearing losses remains an unsolved problem. Early treatment of acute acoustic trauma seems to prevent it partially [*Jakobs and Martin, 1977*].

Although the success rates of any treatment in acute acoustic trauma are limited by theoretical considerations as well as by clinical experience, such an acute inner ear damage has to be considered an emergency which has to be treated as soon as possible. A 'wait and see' attitude denies the patient a chance which should be seized during the very first days after the trauma.

References

Becker, W.; Matzker, J.: Akustischer Unfall. Z. Lar. Rhinol. *40:* 49 (1961).

Birnmeyer, G.: Der akute Lärmschaden. Z. Hörgeräte-Akustik *9:* 102 (1970).

Boenninghaus, H. G.: Ungewöhnliche Form einer Hörstörung nach Lärmeinwirkung und Fehlbelastung der Halswirbelsäule. Z. Lar. Rhinol. *38:* 585 (1959).

Fisch, H.; Murata, K.; Hossli, G.: Measurements of oxygen tension in the human perilymph. Acta oto-lar. *81:* 278 (1976).

Gravendeel, D. W.; Plomp, R.: Micro-noise trauma? Archs Otolar. *71:* 656 (1960).

Herrschaft, H. F.: Die regionale Gehirndurchblutung. Schriftenreihe Neurologie, vol. 15 (Springer, Berlin 1975).

Jakobs, P.; Martin, G.: Die Behandlung knalltraumatischer Innenohrschäden mit Dextran. 40. HNO 25: 349 (1977).

Kawata, S.: Noise-susceptibility, native and acquired. Otol. Fukuoka 13: 1 (1967).

Kellerhals, B.: Acoustic trauma and cochlear microcirculation. Adv. Oto-Rhino-Laryng., vol. 18 (Karger, Basel 1972).

Kellerhals, B.: Medikamentöse Therapie der Schwerhörigkeit. Ther. Umschau 35: 572 (1978).

Lehnhardt, E.: Die Berufsschäden des Ohres. Arch. Ohr.-Nas.-KehlkHeilk. 185: 11 (1965).

Martin, G.; Jakobs, P.: Klinischer Vergleich der Monosubstanzen Dextran 40 und Xantinol-Nicotinat in der Therapie des Knalltraumas. Lar. Rhinol. 56: 860 (1977).

Rüedi, L.; Furrer, W.: Das akustische Trauma. Pract. oto-rhino-lar. 8: 177 (1946).

Spoendlin, H.: Das ischämische Syndrom des Innenohres. Pract. oto-rhino-lar. 31: 257 (1969).

Spoendlin, H.: Akustisches Trauma; in Berendes, Link, Zöllner, Hals-Nasen-Ohrenheilkunde in Praxis und Klinik, vol. 6/II (Thieme, Stuttgart 1980).

Spoendlin, H.; Brun, J. P.: Relation of structural damage to exposure time and intensity in acoustic trauma. Acta oto-lar. 75: 220 (1973).

Prof. B. Kellerhals, Universitäts-HNO-Klinik, Inselspital,
CH-3010 Bern (Switzerland)

Adv. Oto-Rhino-Laryng., vol. 27, pp. 121–129 (Karger, Basel 1981)

Acute Bilateral Hearing Loss During a Pop Concert: Consideration for Differential Diagnosis

H. Irion

Institut für praktische Arbeitsmedizin, Freiburg i. Br., FRG

Introduction

Reports about auditory damage caused by music have been on the rise in the past few years [12]. Rock'n'roll and pop music in particular have been held responsible for auditory damage, whilst this danger appears to be lower in the case of classical music [10]. Not only orchestra and band musicians, audio-engineers and disc jockeys, even pop music listeners are supposedly endangered [16, 17]. Whilst the reports relate in general to chronic sound trauma, the following paper is intended to report a case of acute hearing loss after a visit to a single pop concert.

Case Report

The patient is employed as an administration employee in a factory with industrial noise in which routine auditory examinations are conducted. He had his hearing checked at the factory after visiting the concert and the findings were therefore not made under clinical conditions. A follow-up examination was conducted in a hospital 2 years later.[1]

The patient had undergone otological examination during routine examinations conducted at the large power plant in Mannheim in 1971 and 1977. A pure tone and speech audiogram was made in both instances. Otological examinations and the speech audiogram did not reveal any pathological findings. The pure tone audiogram substantially revealed normal values in addition to minor hearing losses above 1.5 kHz (fig. 1).

---

[1] I am indebted to Priv.-Doz. Dr. *M. Hülse* for conducting the follow-up examinations at the ENT Hospital of the Department for Clinical Medicine in Mannheim.

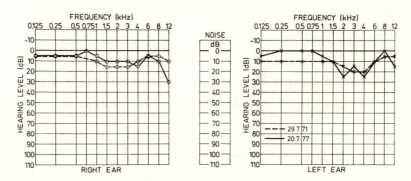

Fig. 1. The pure tone threshold for air conduction before the administration employee visited the pop concert; this examination was performed in the employee's factory. Bone conduction was not plotted unless this was of special importance.

Six months after the last routine examination the 40-year-old employee appeared for an auditory check, since he reported that his sense of hearing had suddenly deteriorated while visiting a pop concert the previous evening.

The concert was held in a hall approximately 50 m in length. A beat band comprising 6 musicians played at this concert. The patient was seated to one side of the gallery about 10 m in front of the loudspeakers. After leaving the concert about 2.5 h later, he realized that his hearing was very poor. He was hardly able to perceive the engine noises of his motor car, for instance. Moreover, he described a high-frequency tinnitus in both ears. His hearing had improved considerably by the next morning.

At auditory examination the morning after the concert, a dip was found bilaterally with a hearing loss of up to 50 dB in the range between 1.5 and 2 kHz. The SISI test was positive bilaterally with 95 and 100% at 2 kHz (fig. 2). The hearing loss initially recovered spontaneously and consequently responded to 'vasodilatatory' treatment on an out-patient basis. The hearing threshold reached the initial values during the course of the next 8 weeks (fig. 2).

At follow-up examination in the hospital on 22 January 1980, the hearing threshold had again elevated slightly (fig. 3). The pure tone and speech audiograms revealed values in the normal range. The slight bilateral tinnitus could be masked at 6 kHz with narrow-band noise of 15 and 20 dB above the hearing threshold. The results of threshold tone decay test (TTDT) according to Carhart were within normal limits. In the noise audiogram according to Langenbeck, the noise threshold converged bilaterally at the reference point. Electronystagmography revealed no pathological findings in the vestibular system.

The case history revealed that the patient had contracted measles and chickenpox as well as a unilateral disorder of the middle ear as a child. In 1957 he suffered a minor brain concussion without losing consciousness and without vomiting. There were no diseases or other factors which could have resulted in auditory damage. The patient was employed in a factory with industrial noise, but being in the administra-

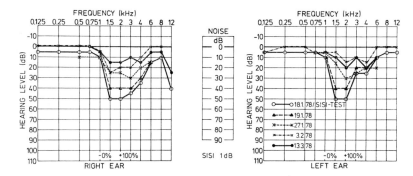

Fig. 2. The pure tone threshold of the patient in figure 1 showing the dips and positive absence of the SISI test at 2 kHz; this diagram was made the morning after the pop concert (18 January 1978). The dip receded during the course of the ensuing 8 weeks.

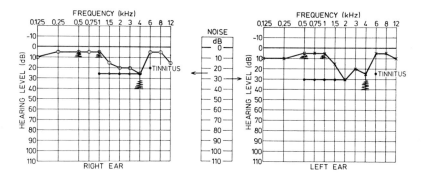

Fig. 3. A follow-up examination of the patient in hospital (22 January 1980). This figure illustrates both the pure tone threshold for air conduction, the results of TTDT according to Carhart at 0.5, 1 and 4 kHz and the noise audiogram according to Langenbeck.

tion he was practically not exposed to noise. The family case history revealed that the patient's mother is hard of hearing. Other cases of impairment of hearing were not known to him.

This case involves a bilateral sensorineural impairment of hearing after short-term noise exposure at a pop concert which recovered in the course of a few weeks. Recovery was probably spontaneous, since the patient's hearing had improved substantially prior to the commencement of treatment.

This case calls to mind a report by *Brusis* [6] relating to an acute loss of hearing with bilateral dips in the case of a 16-year-old girl after a pop concert which lasted 3 h. *Morimitsu* et al. [19] also described an acute loss of hearing with a high-frequency dip after a rock'n'roll concert.

Differential Diagnosis

In acute noise trauma (Schalltrauma) *Feldmann* [8] makes a differentiation between impulse-noise trauma (Knalltrauma), blast trauma (Explosionstrauma), acute inner ear damage by short-term exposure to high intensity noise (Lärmtrauma), and acoustic accident (akustischer Unfall).

The loss of hearing, which occurred in our case immediately after the end of noise exposure and initially receded quickly, speaks in favour of *inner ear damage by short-term exposure to high-intensity noise* [8]. The development of inner ear damage by short-term exposure to high-intensity noise presupposes very high sound levels: 130–160 dB for a few minutes according to *Feldmann* [8]. It was not possible in our case to determine the effective sound level during the concert. Comparison values from the literature [12], however, indicate that the sound level in concerts by beat and rock'n'roll bands does not reach these intensities. A diagnosis of inner ear damage by short-term exposure to high-intensity noise is therefore unlikely, since in our case an adequate sound level did not exist.

What is termed an *acoustic accident* (akustischer Unfall) [2] occurs at lower noise intensities. The prerequisite for an acoustic accident is the abnormal position of the cervical vertebrae which, according to *Boenninghaus* [4], causes the blood supply to the internal ear to be restricted so that irreversible damage can arise if there is simultaneous noise exposure in excess of 90 dB(A). An acoustic accident can be ruled out in our case due to the fact that the hearing loss is bilateral. It may be assumed, however, that abnormal pressure was exerted on the cervical vertebrae, as the patient was sitting to one side of the gallery during the concert. The curve of the hearing threshold with a marked dip and rapid recovery of the hearing loss is not indicative of a diagnosis of a so-called acoustic accident.

In principle, our case could also involve an *acute bilateral essential sudden deafness* in which the noise exposure would only be of secondary importance as the trigger mechanism. This coincidence has been described by numerous authors [2–5, 20, 25]. However, owing to the bilateral aspect of the disorder – a feature rather rare in the case of sudden deafness – together with a marked dip bilaterally in our case, this diagnosis is not entirely convincing either.

What other factors could be of importance in establishing a diagnosis if none of the aforementioned diagnoses appears to apply? The patient's hearing loss certainly cannot be evaluated properly without taking the

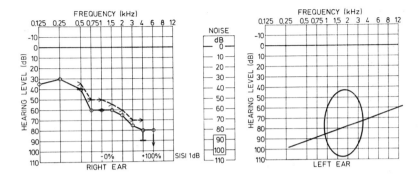

Fig. 4. The audiogram of the mother of the patient in figures 1–3; this audiogram was made in hospital on 5 February 1980. The pure tone threshold on the right includes air and bone conduction, the left ear is deaf. The results of TTDT according to Carhart at 0.5, 1 and 4 kHz are cited on the right.

family case history into consideration, particularly since the mediocochlear dip is indicative of hereditary ear damage. The 70-year-old mother of the patient was hard of hearing. At examination, severe inner ear damage was found on the right side with an absence in the high frequency range and complete deafness on the left (fig. 4). The TTDT according to Carhart revealed values within the normal range. The stapedius reflex did not exist on either side, even at intensities of 130 dB. Electronystagmography revealed no pathological findings. The otoscopic findings were not pathological on either side.

The hearing loss developed after bomb attacks during the war after the mother had repeatedly been deafened for hours or days. During these attacks, she had been in a bunker or cellar in each instance. There was no direct bomb hit. Her husband and other persons who had been subjected to these attacks under similar conditions had not suffered any ear damage themselves. The patient's hearing had deteriorated slightly since the end of the war. The patient was unable to state exactly when her left ear had gone completely deaf. She has been wearing a hearing aid since 1962. No other diseases or other factors could be found which could have damaged her hearing.

The mother's condition involves a sensorineural hearing loss on the right and deafness on the left side. The hearing loss developed following brief noise exposures during the war. A date for the onset of deafness cannot be ascertained. In spite of the conspicuous findings and the descriptions of the occurrences, the mother's disorder does not involve a late sequela following a blast trauma. An impact-noise trauma cannot be assumed either, since adequate peak levels are unlikely to have occurred in view of the circumstances described by the mother. One must assume, however, that the patient's mother also suffered inner ear damage by short-term exposure to high-intensity noise. The severe hearing loss, which did not develop in other persons exposed at the same time she was, as well as the extent of the hearing loss, indicate that the ear reacted inadequately to the sound incidents described.

The extent of the hearing loss suffered by the mother does not correspond to the noise any more than described above in the case with her son in whom a bilateral acute sensorineural hearing loss occurred after visiting a pop concert. An elevated susceptibility to noise on the part of the internal ear must be assumed in both cases.

Discussion

The individual susceptibility to noise differs greatly in cases of noise-induced hearing loss. The author's own investigations [13] have verified the large range of hearing loss after occupational noise exposure as described by *Passchier-Vermeer* [21]. *Melnick* [18] found substantial variation in interindividual noise susceptibility under laboratory conditions for noise exposure lasting 8, 16 and 24 h.

In animal experiments *Spoendlin* [26] observed regular mechanical alterations of the cochlea after a brief exposure period at noise intensities above 130 dB, whereas between 90 and 130 dB delayed metabolic effects exhibit considerable differences from one individual to another. Even according to *Pfander* [22] the damage mechanism in the case of impulse noise is likely to correspond to a mechanical overload, whilst according to *Kellerhals* [14] both mechanical and biochemical factors are of importance in acute acoustic traumata. In cases of long-term noise exposure [7], however, biochemical factors are predominant.

Numerous causes for the individual susceptibility of the inner ear have been discussed: in addition to diseases of the middle ear, pneumatisation inhibition, dysfunctions of the Eustachian tube and inner ear muscles, cranio-cerebral trauma, metabolic and circulatory diseases, as well as a number of infectious diseases, are held responsible as predisposing factors. *Quante* [23] proved that pre-existent toxic damage of the inner ear could culminate in marked noise damage.

In addition to a disposition due to pre-existant diseases, discussion has also centred on a constitutional weakness of the inner ear and on hereditary susceptibility [9]. *Harada* [11] provided proof that in persons working in industrial noise the strain in members from different families caused by noise varies considerably. On occasion there are certainly fluid boundaries between acquired impairment of hearing and the characteristic manifestations of hereditary deafness which are frequently coupled with other abnormalities [15]. *Arnold and Morgenstern* [1] assume that a *genetically induced weakness* and the resultant diminished chances of survival of the hair cells in the inner ear are responsible for the fact that they are damaged by exogenous factors which would not yet cause damage to a normal hair cell. If the destruction of the cell is preceded by an impairment of function of the *stria vascularis* [1], as *Schuknecht* et al [24] described for presbyacusis, this would explain the slow progression and recovery phases even in cases of hereditary deafness.

In our case of acute bilateral inner ear damage after a visit to a pop concert, the mother's deafness, which also occurred after short-term exposures to noise, infers that the cause of the prominent loss of hearing was a *genetically induced weakness of the cochlea* [1].

Summary

A 40-year-old male administration employee suffered an acute acoustic trauma while attending a pop concert. This paper discusses the differential diagnosis. The patient's mother was also hard of hearing following bomb attacks during the last war. The degree of hearing loss did not correlate with the effect of the noise in either case. It is therefore assumed that the inner ear has developed greater susceptibility to noise. The origin of this is attributed to a genetically induced weakness of the cochlea to exogenous factors.

References

1   Arnold, W.; Morgenstern, C.: Hearing impairment in children. Its problems and prognosis. HNO *28:* 10–18 (1980).
2   Becker, W.; Matzker, J.: Akustischer Unfall. Z. Lar. Rhinol. Otol. *40:* 49–58 (1961).
3   Boenninghaus, H. G.: Ungewöhnliche Form der Hörstörung nach Lärmeinwirkung und Fehlbelastung der Halswirbelsäule. Z. Lar. Rhinol. Otol. *38:* 585–592 (1959).

4   Boenninghaus, H. G.: Wann soll eine akute Hörstörung als Arbeitsunfall an-
    erkannt werden? Z. Lar. Rhinol. Otol. *41:* 661–668 (1962).

5   Bosatra, A. B.; De'Stefani, G. B.: The idiopathic sudden deafness. Acta oto-
    lar. suppl. 169, pp. 1–62 (1961).

6   Brusis, T.: Die Lärmschwerhörigkeit und ihre Begutachtung, p. 88 (Demeter,
    Gräfelfing 1978).

7   Drescher, D. G.: A review of general cochlear biochemistry in normal and
    noise-exposed ears; in Henderson, Hamernik, Dosanjh, Mills, Effects of noise
    on hearing, pp. 111–127 (Raven Press, New York 1976).

8   Feldmann, H.: Das Gutachten des Hals-Nasen-Ohren-Arztes, pp. 106–117 (Thie-
    me, Stuttgart 1976).

9   Gestewitz, H. R.: Die adaptive Funktion des Mittelohrapparates zum Schutz
    der Cochlea; Wissenschaftl. Beitr., pp. 222–241, Halle-Wittenberg (1976).

10 Gibbs, G. W.; Hui, H. Y.-T.: A pilot investigation of noise hazards in record-
    ing studios. Ann. occup. Hyg. *16:* 321–327 (1973).

11 Harada, Y.: Personal susceptibility and familial tendency in occupational deaf-
    ness. Auris Nasus Larynx *3:* 31–40 (1976); cited in Zentbl. Hals-Nasen-Ohren-
    heilk. *117:* 517 (1978).

12 Irion, H.: Musik als berufliche Lärmbelastung? pp. 20–27 (Wirtschaftsverlag
    NW, Bremerhaven 1978).

13 Irion, H.; Legler, U.: Audiologische Untersuchungen an Lärmarbeitern zur
    Progredienz der Lärmschwerhörigkeit, p. 116 (Wirtschaftsverlag NW, Wilhelms-
    haven 1975).

14 Kellerhals, B.: Acoustic trauma and cochlear microcirculation. Adv. Oto-Rhino-
    Laryn., vol. 18, pp. 91–168 (Karger, Basel 1972).

15 Kessler, L.; Tymnik, G.; Braun, N.-St.: Hereditäre Hörstörungen, p. 52 (Barth,
    Leipzig 1977).

16 Kowalczuk, H.: 'Big-beat' music and acoustic traumas. Otolaryng. Polska *21:*
    161–167 (1967).

17 Lichtenberg, B.: Zur Frage der Hörschädigung durch Beat-Musik; Diss., pp. 9–
    31, Kiel (1973).

18 Melnick, W.: Human asymptotic threshold shift; in Henderson, Hamernik, Do-
    sanjh, Mills, Effects of noise on hearing, pp. 277–289 (Raven Press, New York
    1976).

19 Morimitsu, T.; Matsumoto, I.; Ochiai, Y.; Komune, S.; Takahashi, M.; Enatsu,
    K.: Consideration of pathology in acoustic trauma based on clinical pictures
    and experimental studies. Otol. Fukuoka *23:* 781–785 (1977); cited in Zentbl.
    Hals-Nasen-Ohrenheilk. *118:* 295 (1978/79).

20 Neveling, R.: Die akute Ertaubung, p. 24 (Kölner Universitäts-Verlag, Köln
    1967).

21 Passchier-Vermeer, W.: Steady-state and fluctuating noise: its effects on the
    hearing of people; in Robinson, Occupational hearing loss, pp. 15–33 (Academic
    Press, London 1971).

22 Pfander, F.: Das Knalltrauma, p. 80 (Springer, Berlin 1975).

23 Quante, M.: Die verstärkte Gefährdung des Hörvermögens im Lärm durch oto-
    toxische Medikamente, pp. 5–82 (Thieme, Stuttgart 1976).

24 Schuknecht, H. F.; Watanuki, K.; Takahashi, T.; Belal, A. A., Jr.; Kimura, R.

S.; Jones, D. D.; Ota, C. Y.: Atrophy of the stria vascularis, a common cause for hearing loss. Laryngoscope, St. Louis *84:* 1777–1821 (1974).

25 Shapiro, S. L.: Deafness following short-term exposure to industrial noise. Ann. Otol., St. Louis *68:* 1170–1181 (1959).

26 Spoendlin, H.: Anatomical changes following various noise exposures; in Henderson, Hamernik, Dosanjh, Mills, Effects of noise on hearing, pp. 69–90 (Raven Press, New York 1976).

Dr. H. Irion, Institut für praktische Arbeitsmedizin,
Elsässerstrasse 2, D-7800 Freiburg i. Br. (FRG)

Adv. Oto-Rhino-Laryng., vol. 27, pp. 130–137 (Karger, Basel 1981)

Deterioration in Hearing Caused by Hearing Aids in Children?

G. Kittel, D. Axmann

Department of Phoniatrics, Otolaryngological Clinic, Erlangen, FRG

Additional hearing impairment caused by high output levels of hearing aids are much more frequently discussed in Anglo-American literature than in the German [1, 4, 5, 8, 10–13, 15, 16, 18].

A majority of authors attribute the deterioration, especially in the case of medium-degree primary hearing impairment, to faulty fitting or wrong application. Since the general guidelines for fitting and checking hearing aids are known, improper procedure do not need to be discussed separately in this paper. Nevertheless, there are circumstances which present problems even to the experienced paedaudiologist, i. e. where auditory efficiency can only be expected with super-power instruments.

Although in German literature the risk of an additional hearing deterioration caused by a hearing aid was in general not deemed to be serious by *Plath* [14], *Biesalski* [3] had drawn attention to a number of cases of hearing deterioration attributable to the effect of hearing aids. He examined the audiograms of hearing-handicapped children during the period of 1967–1975 and found in roughly 35 % of the children progressive hearing loss between 20 and 50 dB which fell almost exclusively within the frequency range from 1,000 to 6,000 Hz. This alarming observation prompted us to put forward the following questions:

(1) Is there a similar progressive nature of hearing disturbances among the pupils of the school for the deaf and hard-of-hearing which is under our medical care?

(2) Is such a progression most likely attributable to the use of hearing aids or is it rather due to an inexorable fate despite the use of hearing aids?

Number of schoolchildren

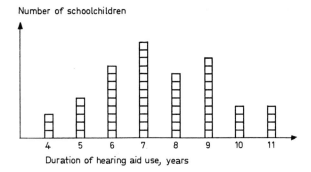

Duration of hearing aid use, years

Fig. 1. Duration of the use of electronic hearing aids.

(3) If changes in the individual dynamic ranges are likely to be caused by high electronic amplification, which is definitely required in the case of residual hearing to obtain auditory efficiency, are we then compelled to refrain from using super-power instruments to restrict the values temporarily, or to employ them only when speech is to be transmitted, or do the sole auditory impressions required for the development of the speech rhythm justify the use of the high sound levels?

In addition to the records on children of the hard-of-hearing classes 6, 7 and 8, we evaluated above all the records of the residual hearing in classes 6, 7, 8 and 9 of the School for the Deaf in Nuremberg. So far, evaluation covered only 55 children, 30 girls and 25 boys, because their hearing thresholds had been determined accurately and the results were repeatedly checked before they had been fitted with hearing aids. Moreover, this test group had been using electroacoustic hearing instruments daily with relatively high output levels over several years (fig. 1).

Since these children had only a very limited speech capability at the time of the first accurate audiometric tests, no useful speech audiograms were available for that period. Therefore, only the old pure tone audiograms were compared with the new audiograms and a measured value tolerance of up to 20 dB was allowed (table I).

11 of the children examined (approximately 20%) revealed a deterioration at one frequency only (table II). Since no deterioration was evident at all other frequencies and since in all instances no deterioration was found in the other aided ear, these 11 cases were not yet analysed further.

Table I. Nature and degree of the hearing impairment in the better ear with binaural fitting in most cases (n = 55)

Nature	Degree	
	combined hearing impairment	sound-perception hearing impairment
Medium to high level	7	18
High level to bordering on deafness	4	26

Table II. Hearing deterioration at one frequency only (n = 11)

Patients	Hearing impairment	
2	below	1,000 Hz
5	between	1,000 and 3,000 Hz
2	in the C₅ range	4,000 Hz
2	In the high-frequency range	> 4,000 Hz

We determined the changes in the individual dynamic ranges with a hearing loss increase from 15 to 50 dB in 7 children (approximately 13 %; fig. 2). The hearing disorders of these 7 children were most likely acquired in 5 cases and hereditary in 2 instances. Hereditary factors were assumed to exist if other members of the family suffered from such disorders, with the exception of presbyacusia. Neither the ENT status nor the impedance tests invariably carried out furnished an indication that any of the 7 children suffered from a middle-ear disease at the time of the last audiometric examination.

The deterioration of the hearing threshold in the children K_1, K_3 and K_5 lies within the frequency-dependent amplification range of the hearing aids. In K_2 and K_4, there is only a partial overlap between these ranges and the threshold shift. The hearing deterioration of K_6 lies outside the amplification range. This child suffered from scarlet fever with otitis media at the time of his hearing threshold deterioration. Furthermore, his brother was also hearing-handicapped. The hearing of child K_7 deteriorated abruptly in the course of an influenza infection. Consequently, the hearing aid cannot be looked upon as the exclusive cause of the hearing deteriora-

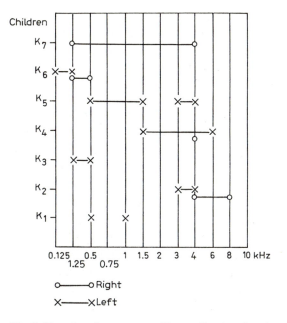

Fig. 2. Deterioration ranges without allowing for the degree of impairment.

tion in K_6 and K_7. As far as K_2 and K_4 are concerned, the hearing aid cannot be regarded as exclusively responsible either, since also frequencies outside the range of influence of the hearing aids used are involved. As the mother of child K_3 also suffers from a rather severe hearing impairment, a hereditary progressive nature of the disease cannot be excluded, irrespective of the burden imposed on the user by the hearing aid. As a result, only in 2 (K_1 and K_5) out of 7 children, hearing-aid produced sound pressures can be primarily seen as the cause of threshold deterioration.

In K_1, however, the hearing deterioration was observed only in one ear even though the same hearing aid was worn in both. Consequently, only in child K_5 the hearing deterioration is most likely due to the use of a hearing aid although earlier experiences have already demonstrated that especially very severe sensorineural damage in the early childhood may become spontaneously progressive, independent of the use of hearing aid, even when no signs exist suggesting a family history.

The studies which involved 55 children using hearing aids for years and checked at regular intervals did not confirm the existence of negative, instrument-related sound influences on the individual dynamic ranges. Even in 7 children with clear threshold deterioration almost invariably more likely causes were found to be at least contributary to the progressive nature of the deterioration in hearing.

In children with a relatively high sensitivity to sound volumes the maximum output values were always brought to below the threshold of discomfort by peak clipping or automatic volume control. Although the children examined furnished no conclusive evidence of instrument-caused threshold deterioration, we would nevertheless draw attention to the possibility of hearing-aid-induced damage in the case of high or relatively excessive output levels, especially when deteriorating factors, independent of the hearing aid, are present. This possibility should also be considered when the deterioration develops in combination with intercurrent otic diseases or in the case of increased vulnerability. As experimental investigations showed that noise-induced damage develops more readily with concurrent oxygen deficiency, an additional instrument-induced damage should not be ruled out even in a patient with a family history or intercurrent diseases, although the contribution made by the various factors is difficult to assess.

In the light of the present investigations it can be assumed that under roughly normal starting conditions an impaired ear is hardly likely to be additionally damaged by a correctly fitted hearing aid and regular setting checks; this assumption holds also good when relatively high output levels are applied.

However, the use of a hearing aid should be prohibited when the hearing-handicapped person suffers from a recognizable intercurrent ear disease. The question as to the consequences in the case of residual hearing and hereditary weakness is more difficult to answer. The use of accurately fitted and regularly checked hearing aids should not even be denied to such children for fear of hearing-aid-induced threshold deteriorations, although *Kinney* [6] pointed out that persons with a hereditary history appeared to be most likely to become victims of hearing-aid-induced deteriorations. However, in the case of children with a hereditary predisposition, one must expect that, despite the decision to dispense with a hearing aid, the last hearing residues will also be lost as a result of hereditary progression, without speech impressions ever having been imparted. Preserving residual hearing would be meaningless if it is not utilized for speech development which in turn is not possible without high amplification when severe hearing impairment is present.

Although even in early childhood a spontaneous progressive deterioration can frequently be observed, the fitting of severely hearing-impaired children or those with residual hearing with a hearing aid was occasionally considered to be convenient at a later date, as additional instrument-produced auditory deterioration was feared. The underlying reason for this approach was that the speech centres would be given more time to mature in their development independent of the speech input, and that the children would acquire an improved efficiency in learning during the periods the hearing aid is worn. We do not concur with this view because the investigations of our patients suggest that precisely during the first 2 years of life the control of acoustic attention is of particular importance to the development of speech.

The compromise of accepting a somewhat protracted delay in fitting such children with hearing aids would, in addition, put up with a hearing-aid-independent progression without full exploitation of still better hearing phases. We know of children whose parents rejected the use of high-gain hearing aids, who in spite of that lost their hearing so that they were unable to benefit from acoustic aids in their development of speech. It should be noted, however, that, if the paedaudiologist had insisted on the use of hearing aids, he would have had to reckon with a regression.

Hearing aids should be fitted accurately and as early as possible in the case of high-grade hearing impairment and even sensorineural hearing loss bordering on deafness, because only in this manner it is be possible to utilize the best acoustic phases of attention. In such cases, the gain appears to be higher than the risk of instrument-induced injuries which the present investigations showed to be not very serious. With such severe hearing impairments, the most sensitive sensory cells, which would be damaged by such volumes in the case of normal hearing, would no longer function properly, and only sensory cells with a high stimulation threshold would respond. Hearing aids should be worn all day long if the daily offer of acoustic stimuli is normal; if the burden of acoustic stimuli is above normal, the aids should be worn only for shorter periods to allow still functioning sensory and nerve cells to regenerate especially when ears respond positively to recruitment. Regular audiometric checks and hearing-aid inspections are now as ever mandatory to avoid additional acoustic damage. If the well-known guidelines are observed, the risk of instrument-related deteriorations is so slight – this was confirmed by the present investigations – that the children should not be denied the benefits of such aids.

Summary

(1) There were 7 children among 55 afflicted with severe hearing impairment or with residual hearing (approximately 13 %) who exhibited mostly unilateral progression which was ascertained by accurate pure tone audiometric tests during an observation period of several years.

(2) Instrument-caused deterioration, however, could not be confirmed in these children. Except for 1 case, deterioration was rather attributable to intercurrent diseases and hearing-aid-independent progression. However, a higher sensitivity to additional auditory damage appears to exist.

(3) According to the results of the present studies, the use of super-power hearing aids is indicated in cases of residual hearing although the hearing aid should not be used during the period of intercurrent diseases. In early childhood, intensive speech hearing training should in any case be limited to about three periods of 20 min each daily. This also applies when speech is not understood because experience demonstrates that the acquisition of speech rhythms is a major aid in the process of learning to speak.

References

1 Barr, B.; Wedenberg, E.: Prognosis of perceptive hearing loss in children with respect to genesis and use of hearing aid. Acta oto-lar. *59:* 462 (1965).

2 Bellefleur, P. A.; Van Dyke, R. C.: The effect of high gain amplification on children in a residential school for the deaf. J. Speech Hear. Res. *11:* 343 (1968).

3 Biesalski, P.: The rehabilitation of the child with hearing impairment. Z. Hörgeschädigtenpädagogik *1:* 1 (1977).

4 Jerger, J. F.; Levis, N.: Binaural hearing aids: Are they dangerous for children? Archs Otolar. *101:* 480 (1975).

5 Kinney, E.: Hearing impairment in children. Laryngoscope, St Louis *63:* 220 (1953).

6 Kinney, C. E.: The further destruction of partially deafened children's hearing by use of powerful hearing aids. Ann. Otol. *70:* 828 (1961).

7 Macrae, J. H.: Deterioration of the residual hearing of children with sensory neural deafness. Acta otolar. *66:* 33 (1968).

8 Macrae, J. H.; Farrant, R. H.: The effect of hearing aid use on the residual hearing of children with sensory-neural deafness. Ann. Otol. *74:* 409 (1965).

9 Maran, A.: The cause of deafness in childhood. J. Laryn. *80:* 495 (1966).

10 Markides, A.: Do hearing aids damage the user's residual hearing? Sound *5:* 99 (1971).

11 Møller, T. T.; Rosjkjaer, C.: Injury to hearing through hearing air treatment. 5th Congr. of Int. Soc. of Audiology, Bonn 1960.

12 Murray, N.: Hearing aids and classification of deaf children. Report C. A. L. -1R-2 Commonwealth Acoustic Laboratories, Sydney 1951; cited by Macrae and Farrant.

13 Naunton, R.: The effect of hearing aid use upon the user's residual hearing. Laryngoscope, St Louis *67:* 569 (1957).
14 Plath, P.: Das Hörorgan und seine Funktion (The hearing organ and how it functions) (Carl Marhold Verlag, Berlin 1976).
15 Rice, J. C.: Hearing aids. J. Laryng. *80:* 583 (1966).
16 Roberts, C.: Can hearing aids damage hearing? Acta oto-lar. *69:* 123 (1960).
17 Ross, M.; Lerman, J.: Hearing aid usage and its effect upon residual hearing. Archs Otolar. *86:* 639 (1967).
18 Ross, M.; Truex, E. H.: Protecting residual hearing in hearing aid user. Archs Otolaryn. *82:* 615 (1965).

Prof. Dr. med. G. Kittel, Phoniatrische Abteilung der Universitäts-HNO-Klinik, Waldstrasse 1, D-8520 Erlangen (FRG)

Adv. Oto-Rhino-Laryng., vol. 27, pp. 138–143 (Karger, Basel 1981)

Inner Ear Hearing Loss in Acute and Chronic Otitis media

G. Münker

ENT Department of the University Hospital of Freiburg/Breisgau, FRG

The topic of sudden hearing loss should include such conditions which are often observed in chronic otitis media when, for example, a cholesteatoma has led to fistulation of the labyrinth: At first a circumscript labyrinthitis develops, sooner or later a diffuse purulent labyrinthitis with a accompanying loss of inner ear function evolves. A labyrinthine fistula will often lead to, but it does not so necessarily, complete deafness.

The grave and often life-threatening complications following middle ear inflammations often lead to complete inner ear deafness, whereas the hearing impairment that results from residuals of middle ear disease on the tympanic membrane and the ossicles is the classical form of sound conduction hearing loss. An accompanying inner ear hearing loss in cases of acute and chronic otitis media is, although on the daily order of events, hardly mentioned in textbooks and in the literature. Among the various possibilities leading to inner ear hearing loss the inflammatory middle ear diseases take a central role: According to *Morrison* [1969] middle ear inflammation can be held responsible for about one fourth of cases exhibiting serious hearing loss of the inner ear.

With advancing age hearing loss in the affected ear progresses with respect to the normal ear [*Paparella and Brady,* 1970]. This makes prompt operative treatment essential. The impairment of bone conduction increases with the duration of the illness [*Paparella and Brady,* 1970]. Statistical investigations show cochlear damage to be present in almost half of those patients with histories of middle ear suppuration over the course of many years [*Gardenghi,* 1955; *Bluvshtein,* 1963].

We are very familiar with audiograms showing combined hearing impairment in chronic otitis media. We usually find reductions in bone conduction toward the higher frequencies in cases of acute otitis media, bullous myringitis and otitis media with effusion. Cases of otitis media with effusion show a reduction of bone conduction toward the high frequencies in about 40 % [*Münker, 1977*], surprisingly more often in serous than in mucoid effusions. Usually hearing can be restored to normal after successful ventilation of the middle ear. This is important to know, in order not to assume permanent inner ear damage, which is the case after longstanding chronic otitis media.

One glance at the anatomical relationship between the middle and the inner ear suffices to explain how diseases of the middle ear can lead to inner ear affection across the round window. With the available evidence, we cannot but accept two potential routes from the middle ear to the inner ear: round window membrane and lymphatics [*Goycoolea* et al., 1980; *Arnold and von Ilberg, 1972*]. The following possibilities should be considered as causes of inner ear involvement in cases of acute or chronic otitis media:

(1) If there is reversible loss of bone conduction one should consider the possibility that liquid and mucous have constricted the movement of the round window membrane. Impairment of bone conduction and reduction of cochlear potentials resulting from plugging of the round window has been experimentally demonstrated by *Tonndorf and Tabor* [1962], *Gisselson and Richter* [1955], *Wever* et al. [1948] as well as *Ranke* et al. [1952]. After ventilation of the middle ear, air and bone conduction return to normal.

(2) Vibrations of the cranium are conveyed also into the air-containing spaces of the skull. This so-called secondary aerial sound contributes to bone conduction. When the middle ear is filled with fluid, this form of osteotympanic sound transmission is hindered; thus, we have a further explanation for reversible impairment of bone conduction.

(3) The round window membrane plays a central role in the discussion of inner ear involvement in otitis media. This membrane consists of three layers: an epithelial layer, composed of flat squamous cells, an inner layer of elongated thin cells with thin cytoplasmatic extensions serving as the immediate barrier to the perilymph, and between these a middle mesenchymal layer forming a fibrocytic 'sponge' with collagen and elastic fibers filled with fluid. It is across these sponge pores that lymph fluid in the middle ear mucosa is able to freely communicate with the perilymph [*Arnold and von Ilberg, 1972*]. Knowledge of such direct communication between the middle and inner ear was the basis upon which pharmaco-

therapy of the inner ear across the middle ear was developed, as exemplified by the streptomycin and gentamycin therapy in Menière's disease. The permeability of the round window is increased under conditions of inflammation accompanied by pH fall. More recent investigations suggest that noxious elements may pass through the round window and cause inner ear pathologic changes. The passage of electrolytes (Na^{22}), horseradish peroxidase, macromolecules (tritiated albumin), and isolated bacterial toxins (staphylotoxin) has been observed diffusing across the round window membrane [*Paparella* et al., 1980]. Blockage of lymph drainage in the subepithelial zone as seen in the course of tissue transformation in inflammation can lead to a damming of end products of metabolism in the tissue and to crossing of such end products into the perilymph.

(4) After experimentally inducing middle ear inflammation in the guinea pig *Rauch* [1965] observed an immense increase in the perilymph content of such enzymes which are increased in the mucosa in chronic middle ear affections and otitis media with effusion as observed by *Palva* et al. [1974] (e. g. lactate dehydrogenase). *Hache* et al. [1969] found immense increases in the whole protein concentration of the perilymph following experimentally induced middle ear inflammation.

Histologically, one finds serofibrinous precipitates with and without accompanying inflammatory cells in the perilymph of the scala tympani of the basal turn of the cochlea. Rarely swelling and detachment of the tectorial membrane as well as damage to the outer hair cells can be seen. What we have here is the histological picture of a serous labyrinthitis, typically found in the immediate vicinity of the round window and limited to the basal turn of the cochlea. *Paparella* et al. [1972] found these serofibrinous precipitates within the perilymph to be present in almost all cases of otitis media, yet in only 3 % of the normal petrous bones which were examined. Thus, we must assume from this that contamination of the perilymph and subsequent labyrinthitis is present in every case of acute otitis media. Although the damage is usually reversible, it is possible that permanent damage to the inner ear can result as the aftermath of recurrent middle ear inflammation reaching the end organ in the long run. Perhaps threshold shifts beyond the cognitive threshold appear with every bout of otitis media. These changes may remain unrecognized during childhood, and we do not know whether or not this makes the affected ear more susceptible to prebyacusis or noise-induced hearing loss, or if there is an additive effect involved.

(5) *Paparella* et al. [1980] observed the coincidence of otitis media and endolymphatic hydrops in a temporal bone study. Endolymphatic hydrops as a complication of otitis media may explain fluctuant sensorineural hearing loss.

(6) It is precisely the zone surrounding the round window niche that reveals numerous vascular channels between the middle ear mucosa and the mesenchymal lining of the perilymph space. Typically witnessed changes in chronic inflammatory states such as circulatory disturbances, vasoconstriction and vasodilatation of mucosal vessels could influence the inner ear along this path. It is also possible that toxic products of metabolism reach the internal ear by way of the perivascular space.

(7) Finally, the possibility that chronic mucosal secretions and the accumulation of mucous exudates may hinder the free diffusion of oxygen into the inner ear must be discussed. *Maas* et al. [1976] showed in animal experiments that the oxygen tension of the perilymph is directly proportional to the oxygen tension in the middle ear. Normally the oxygen tension in the perilymph behind the round window is higher than in the area of the organ of Corti, yet lower than that of the middle ear. The investigatons of *Morgenstern* [1980] showed that the oxygen tension in the middle ear mucosa falls parallel to that in the perilymph when the oxygen partial pressure in the middle ear is lowered or when nitrogen respiration is conducted.

It is important to mention also the possibility of inner ear damage following the *therapy* of middle ear affections, especially when considering the ototoxic antibiotics. Even the attempts to correct residual disease following otitis media operatively may lead to an impairment of bone conduction threshold. *Bartual* [1964] found postoperatively a reduction in bone conduction in a high percentage of cases. The postoperative damage affected one or both functions of the inner ear; isolated cochlear damage was witnessed in only 5% and isolated vestibular damage in 35%, and damage of both functions in 17%. Since vestibular damage is more easily recognized by ENG inner ear hearing loss can be anticipated; on the other hand, this also disappears earlier and thus would point toward recovery of the cochlear function.

It is a surprising fact that a reduction in bone conduction can be witnessed quite often after a type I tympanoplasty, after the mere closure of a tympanic membrane perforation. Only a portion of these cases showed normalization or improvement later [*Beck and Münker,* 1974]. This phenomenon appears to be an expression of underlying inflammation or toxin invasion due to operative trauma. The pathogenesis of bone conduction loss would be similar to that described for serous labyrinthitis.

The findings presented here serve to show that in inflammatory ear diseases also the function of the inner ear should be given more attention. Data from patients and animal experiments clearly demonstrate that the middle and inner ear must be looked upon not only anatomically but also functionally as one unit.

Summary

Temporary or permanent threshold shift of bone conduction occurs frequently in acute or chronic otitis media. The sensorineural hearing loss is dependent on the age of the patient and the duration of the illness. Fluid in the middle ear may impede sound transmission and oxygen transport to the inner ear. In middle ear inflammation noxious substances may pass across the round window membrane leading to serous labyrinthitis. In therapy ototoxic drugs and operations (tympanoplasty) can cause sensorineural deafness.

References

Arnold, W.; Ilberg, C. von: Neue Aspekte zur Morphologie und Funktion des runden Fensters. Lar. Rhinol. *51:* 390–399 (1972).

Bartual, J.: Die Innenohrreaktion nach der Tympanoplastik. Arch. Ohr.-Nas.-Kehlk-Heilk. *184:* 151–165 (1964).

Beck, C.; Münker, G.: Schädigung des Innenohres nach Tympanoplastik Typ I. Archs Otol. *78:* 371–375 (1974).

Bluvshtein, G. M.: Audiologicheskaia, Kharakteristika, khronicheskikh guoinykh sreduikh otitov. Vest. Otorinolar. *25:* 64 (1963).

Gardenghi, G.: Contribute allo studio della funzione cochleare nell'otite media purulenta chronica. Boll. Mal. Orecch. *73:* 587 (1955).

Gisselson, L.; Richter, N.: Ein Beitrag zur Frage des Hörvermögens bei Verschluss des runden Fensters. Arch. Ohr.-Nas.-KehlkHeilk. *166:* 410–418 (1955).

Goycoolea, M. V.; Paparella, M. M.; Juhn, S. K.; Caspenter, A. M.: Oval and round window changes in otitis media. Potential pathways between middle and inner ear. Laryngoscope, St Louis *90:* 1387–1391 (1980).

Hache, U.; Gerhardt, H. J.; Scheibe, F.; Haupt, M.; Ritter, J.; Rabenow, M.: Otitis media und Kochlea: Morphologische und biochemische Untersuchungen am Meerschweinchen. Archs Oto-Rhino-Lar. *214:* 49–61 (1969).

Maas, B.; Baumgaartel, H.; Lübbers, D. M.: Lokale pO_2- und pH_2-Messungen mit Nadelelektroden zum Studium der Sauerstoffversorgung und Mikrozirkulation des Innenohres. Arch. Ohr.-Nas.-KehlkHeilk. *213:* 439–440 (1976).

Morgenstern, R.: Oxygen supply of middle ear mucosa under normal condition and after Eustachian tube occlusion. Annls Oto-lar., St Louis *68:* suppl., pp. 76–78 (1980).

Morrison, A. W.: Management of severe deafness in adults. Proc. R. Soc. Med. *62:* 959–965 (1969).

Münker, G.: Knochenleitungsveränderungen beim Sero-Mucotympanon. Lar. Rhinol. *56:* 591–594 (1977).

Palva, T.; Raunio, V.; Nousianen, R.: Secretory otitis media: protein and enzyme analyses. Ann. *83:* suppl. 11, pp. 35–43 (1974).

Paparella, M. M.; Brady, D.: Sensorineural hearing loss in chronic otitis media and mastoiditis. Trans. Ann. Acad. Ophthal. Oto-lar. *74:* 108–115 (1970).

Paparella, M. M.; Goycoolea, M. V.; Meyerhoff, W. L.: Inner ear pathology and otitis media. Annls Oto-lar., St Louis *68:* suppl., pp. 249–253 (1980).

Paparella, M. M.; Oda, M.; Hiraide, F.; Brady, D.: Pathology of sensorineurol hearing loss in otitis media. Ann. *81:* 632–647 (1972).

Ranke, O. F.; Keidel, W. D.; Weschke, H. G.: Cochleaeffekt bei Verschluss des runden Fensters; in Grützmacher, Meyer, Akustische Beihefte, vol. 3, pp. 146–148 (1952).

Rauch, S.: Die differentialdiagnostische Bedeutung der LDH-Iso-enzyme bei Innenohrerkrankungen. Practica oto-rhino-lar. *27:* 143–147 (1965).

Tonndorf, J.; Tabor, J.: Closure of the cochlear windows. Its effect upon air and bone conduction. Annls Oto-lar., St Louis *71:* 5–29 (1962).

Wever, E. G.; Lawrence, M.; Smith, K.: The effect of negative air-pressure in the middle ear. Annls Oto-lar., St Louis *57:* 418 (1948).

Prof. Dr. G. Münker, ENT Department of the University of Freiburg, Killianstrasse 5, D-7800 Freiburg/Br. (FRG)

Adv. Oto-Rhino-Laryng., vol. 27, pp. 144–158 (Karger, Basel 1981)

Drug-Induced Sudden Hearing Loss and Vestibular Disturbances

P. Federspil

Hals-Nasen-Ohren-Universitätsklinik, Homburg/Saar, FRG

Introduction

Drug-induced cochlear and vestibular disturbances appear as a rule suddenly. 9 cases out of 10 of ototoxic damage are due to aminoglycoside antibiotics or their combination with other ototoxic or nephrotoxic drugs. It is the aim of this report to describe the signs and symptoms of ototoxic damage and in addition to peruse the data collected hitherto in experimental and clinical set-ups concerning the factors influencing the ototoxicity of the aminoglycoside antibiotics and the treatment of ototoxic damage. Nevertheless, we should notice that drug-induced disturbances of the cochlear and vestibular functions were known in the past century, but today the number of possible ototoxic substances is high. Besides the aminoglycoside antibiotics we mention the salicylate analgesics, the loop-inhibiting diuretics, some cancer chemotherapeutic agents, antimalarial drugs as well as the tricyclic antidepressants.

Clinical Symptoms and Signs Due to Ototoxic Lesions

The clinical symptoms and signs due to lesions caused by the new generation aminoglycoside antibiotics resemble those due to lesions developed after the application of the 'classical' aminoglycoside antibiotics. Streptomycin, gentamicin, tobramycin, sisomicin, dibekacin, and netilmicin mainly damage the organ of balance ($2/3$ of the total number of cases observed), whereas kanamycin and amikacin mainly affect the hearing function ($5/6$ of the cases). Vestibular complaints range from a feeling of drunkenness, nausea, vomiting, unsteadiness of the gait – particularly in

the dark, on an uneven surface – swaying, positional and positioning vertigo with or without a nystagmus to complete labyrinth paresis and the
Dandy syndrome. Lack of caloric excitability but perseverance of galvanic response are signs of the peripheral nature of the disorder.

The loss of hearing presents as a pure perception disorder, whereby
the high frequencies are affected first and foremost. However, some cases
show major loss of the low frequencies as in Menière's disease. In general
the recruitment signs are present in agreement with the primary nature of
peripheral damage to the auditory organ, as recognized since *Caussé* [8].
Other cases have been reported where understanding of speech is impaired more than the reduction of the pure tone threshold resulting from
streptomycin, dihydrostreptomycin, and gentamicin therapy.

In those fortunate cases where tinnitus or complaints of pressure in
the ear as well as a sensation of a blocked ear are the first complicating
symptoms, one should consider the discontinuation of the aminoglycoside
antibiotic therapy. Naturally, all the hearing and vestibular disturbances
occurring in everyday life, i. e. otitis media, must also be taken into consideration when discussing the ototoxicity of drugs.

As a rule ototoxic lesions affect both ears to the same extent. Exceptions were already described in streptomycin treatment. *Meyers* [37],
Jackson and Arcieri [32] and *Federspil* [13] claimed a high ratio of unilateral cochlear impairment caused by gentamicin after it was introduced.
However, our experimental work concerning the pharmacokinetics of
gentamicin in the inner ear and the hair cell damage of the organ of Corti
of the guinea pig revealed significant asymmetry in approximately 15 %
of the animals only. Hence we should continue to assume that gentamicin
and the new generation aminoglycoside antibiotics produce symmetrical
lesions as published up to now. This can be substantiated by results we
have obtained subsequently in gentamicin and other aminoglycoside antibiotic tests. It must be noted, though, that early inner ear damage presents
more often unilaterally than advanced damage. A bilateral aminoglycoside
antibiotic hearing loss is more easily distinguished from an idiopathic sudden hearing loss than a unilateral one.

Only a few reports have mentioned the reversability of streptomycin
and kanamycin damage; nonetheless, this is important. In 1970 we first
pointed out the possibility in a case of unilateral hearing loss which had
been verified audiometrically and was caused by gentamicin treatment
[12]. Since then we have managed to demonstrate this feature experimentally in the guinea pig. This applies to gentamicin, tobramycin, siso-

micin, and amikacin damage. Furthermore, we managed to show that the vestibular organ, damaged by the same drug, can also recover. In the meantime clinical reports concerning the late generation aminoglycoside antibiotics have been published in support of our experimental results. *Jackson and Arcieri* [32] and *Federspil* [13] have stated in earlier publications that in approximately half of the cases that developed ototoxic signs following gentamicin treatment, the auditory and vestibular functions recovered. Revival was more common in those cases in which either the treatment was discontinued or at least the dose reduced at the onset of the first clinical signs of ototoxicity. As regards the prognosis of the damage of the vestibular organ, it can be judged as less serious because the loss of its function can mostly be compensated for by the cutaneous and proprioceptive sensory organs as well as the sense of vision.

The question of revival following cochlear and vestibular impairment due to aminoglycoside antibiotic treatment leads to the question of progression and late ototoxicity. Late ototoxicity often encountered after dihydrostreptomycin and neomycin treatment is rarely met after kanamycin and very rarely seen after streptomycin application. Up to now only sporadic cases of late ototoxicity caused by the new generation aminoglycoside antibiotics have been reported except in case of renal disorders. This is confirmed by our follow-up study dealing with 96 patients treated with gentamicin and who were controlled over a period of 1–3 years [12]. None of these showed any definite symptoms or signs of late ototoxicity. However, clinical investigations concerning late ototoxicity need not to be absolutely relevant because the doses applied are often significantly below the ototoxic level and because an attack of sudden hearing loss is difficult to exclude. It is for these reasons that we have developed an appropriate experimental set-up [24]. The results of it show that when no renal disorder was present, no signs of late ototoxicity following gentamicin, tobramycin, and sisomicin application could be found. More recent experiments with amikacin show similar results and were confirmed by clinical observations of 20 patients having been treated with total doses of 171 mg amikacin per kg body weight 3–15 months before [25].

Factors Influencing Ototoxicity

Ototoxic damage due to parenteral administration of aminoglycoside antibiotics depends on several factors.

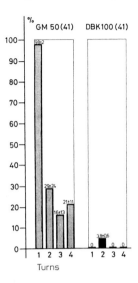

Fig. 1. Histocochleogram after 41 days of gentamicin (GM) at 50 mg/kg/day and dibekacin (DBK) at 100 mg/kg/day.

Intrinsic Ototoxic Potential

The ototoxic potentials of the different aminoglycoside antibiotics are not the same. The importance of well-designed and meticulously executed experimental studies for the evaluation of the ototoxicity of the new aminoglycoside antibiotics, and the pitfalls as well as the clinical relevance of these studies have been pointed out recently [26]. Comparative studies of the ototoxicity of the new aminoglycoside antibiotics considering the nephrotoxic potential were published on gentamicin [14, 16], tobramycin [16, 21], sisomicin [18], amikacin [18, 25], netilmicin [20], and dibekacin [28]. When interpreting these results, attention must be paid to the main general factor of ototoxicity which is the total dose of the aminoglycoside anibiotic administered. Tobramycin is less ototoxic than gentamicin, but in patients with normal renal function and without other risk factors this difference is not relevant when applying the usual clinical dosage. The difference between the ototoxic potentials of sisomicin and gentamicin is even less. On a weight-for-weight basis the cochlear ototoxicity of amikacin is more than half that of gentamicin, whereas the ototoxicity of dibekacin is less than half that of gentamicin (fig. 1). Figure 1 gives a graphic

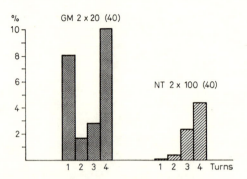

Fig. 2. Histocochleogram after 40 days of gentamicin (GM) at 2×20 mg/kg/day and netilmicin (NTM) at 2×100 mg/kg/day.

representation of the average histocochleograms after 50 mg gentamicin and 100 mg dibekacin per kg body weight per day administered subcutaneously during 41 days to two groups of 4 guinea pigs each. The cochlear and vestibular ototoxicity of netilmicin is less than one fifth that of gentamicin. Figure 2 shows the average percentages of degenerated outer hair cells in two groups of 8 guinea pigs each having been injected subcutaneously 2×20 mg gentamicin and 2×100 mg netilmicin per kg body weight during 40 days. The cochlear as well as the vestibular damage induced by the 5-times higher netilmicin dosage is less than that of the gentamicin group.

It seems best to indicate the relative ototoxic potential of the different aminoglycoside antibiotics first on a weight-for-weight basis. Besides this it is advisable to take the therapeutic dosage into account. If we consider the average clinical dosage of the different aminoglycoside antibiotics, netilmicin is the least cochleotoxic, followed by dibekacin, tobramycin, gentamicin, sisomicin, and amikacin.

Dosage and Route of Administration

Functional and histological investigations of gentamicin, tobramycin, sisomicin, amikacin, dibekacin, ribostamycin, and netilmicin showed that the essential parameter for the evaluation of the ototoxic potential of an aminoglycoside antibiotic – renal impairment excluded – is the total dose of the aminoglycoside antibiotic administered. That is why the upper limit of the daily aminoglycoside dose in relation to the ototoxicity depends essentially on its influence on the total threshold dose. Apart from nephro-toxicity, general toxicity, and therapeutic need must be taken into consideration as well.

In histological investigations with tobramycin and amikacin, we also found that the ototoxicity is not diminished by a division of the daily dose, but even slightly increased. That is the reason why it is probably better to administer the necessary daily dose in two single doses instead of three, which does not seem to reduce the efficacy. These results coincide with those obtained in former investigations by *Tompsett* [46] on streptomycin. The studies on the nephrotoxicity of gentamicin, tobramycin, and netilmicin in dogs performed by *Thompson* [45] showed a lower degree of nephrotoxicity after bolus injection than after continuous intravenous infusion. The results concerning the ototoxicity may be considered as a confirmation of the pharmacokinetic investigations according to which the actual aminoglycoside antibiotic concentrations in the perilymph as well as in serum depend linearly on the dose administered [14, 22, 42], and speak against the existence of a critical serum level, at least in the case of gentamicin and tobramycin [22]. Despite this, the so-called critical serum concentrations of aminoglycoside antibiotics are still mentioned in publications, although for gentamicin already very different amounts of the critical serum level have been indicated: *Jackson and Arcieri* [32] 12 mg/l; *Sweedler* et al. [14] 13 mg/l; *Höffler* [35] 6.8 mg/l a long time ago. *Von Oldershausen* et al. [40] as well as other authors failed to see ototoxic lesions in cases of serum concentrations ranging from 10 to 20 mg/l. In man a critical serum level is likely to be assumed if the inner ear damage was only to be reached above a certain serum level, i. e. if the passage of aminoglycoside antibiotics from the serum to the perilymph were not to take place linearly but only above a certain threshold. In this case, the aminoglycoside antibiotics would have to be considered as very dangerous drugs and the efforts of research as well as of clinical workers to induce a significant cochlear damage in animals or man without renal disturbances could not be explained. The above considerations led us to think that in patients with normal renal function who are able to give information on vertigo and hearing disturbances and not undergoing a high-dose therapy it does not seem necessary to evaluate the serum concentrations of well-known aminoglycoside antibiotics with low ototoxic potential in order to avoid ototoxic damage and to optimate therapy. The importance of serum concentrations for therapy is not so clear as it might appear and the relationship between therapeutic effectiveness and daily

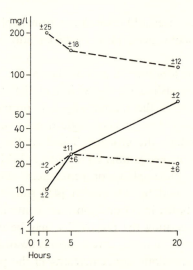

Fig. 3. Dibekacin pharmacokinetics in perilymph (—), serum (– – –), and aqueous humor (– · –) after a subcutaneous injection of 50 mg/kg in bilaterally nephrectomized guinea pigs.

dosage (body lean mass) is probably closer than the relationship between effectiveness and serum peak or through concentrations. The main reason for evaluation aminoglycoside antibiotic serum through concentrations is the fact that this is the most sensitive method for an early detection of functional renal disturbances and, as will be mentioned below, the renal function has a major influence on the ototoxic potential of the amino-glycoside antibiotics [22, 33].

The importance of a former aminoglycoside antibiotic treatment for the ototoxic potential of a new aminoglycoside antibiotic therapy is not well known. It is likely that the total dose of aminoglycoside antibiotics that had been applied some longer time ago cannot be totally added to the newly administered doses of aminoglycoside antibiotics [23].

Renal Function

Since the end of the 1940s, the influence of renal functional disturbances on the ototoxicity of aminoglycoside antibiotics has been known. Figure 3 shows the very highly increased dibekacin concentrations in the perilymph besides the high serum and aqueous humor concentrations in

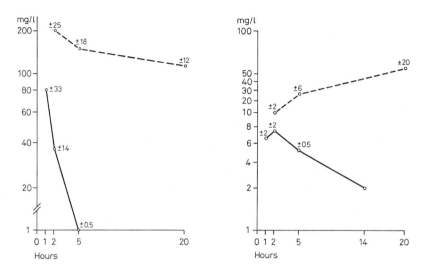

Fig. 4. Pharmacokinetics of dibekacin in serum (a) and perilymph (b) after a subcutaneous injection of 50 mg/kg in the normal guinea pig (—) and after bilateral nephrectomy (– – –).

bilaterally nephrectomized guinea pigs in comparison to normal guinea pigs having got the same subcutaneous injection of 50 mg/kg (fig. 4). These pharmacokinetics of dibekacin resemble those of gentamicin [14, 22, 24]. The most striking phenomenon of the comparative aminoglycoside concentration studies in the bilaterally nephrectomized guinea pigs is the longer retention in the perilymph than in serum and other fluids. Anuria of the guinea pig has pharmacokinetically the same effect on the aminoglycoside antibiotic inner ear concentrations as a 20- to 50-fold elevation of the dosage. As we consider today the inner ear concentrations of the aminoglycoside antibiotics as graduator for the ototoxicity, the well-known influence of uremia on the aminoglycoside antibiotic ototoxicity is experimentally proved by this and quantified. The model of the pharmacokinetics of dibekacin and gentamicin in serum and perilymph of the nephrectomized guinea pig may also be applied for the evaluation of the ototoxicity of permanently high aminoglycoside antibiotic levels in the serum. This mode of administration of aminoglycoside antibiotics recently introduced in the United States of America by *Bodey* et al. [3] does probably not increase their efficacy to the same extent as it increases the ototoxicity [14, 22].

Preexisting Hearing Disturbances

If a cochlear degeneration is already existing, i. e. in case of Menière's disease, syndrome of the cervical spine, cranial trauma, a hereditary inner ear dysacusis, a toxic lesion of another origin or a noise-induced hearing impairment, it is generally presumed that the sensitivity for a toxic lesion is increased, even if this could not be verified in the latest clinical investigations [14]. This problem arose mostly in the case of elderly patients with their presbyacusis. The main danger for elderly patients treated with aminoglycoside antibiotics however lies in neglecting the frequently existing renal functional impairment. It is a fact that the aminoglycoside antibiotics are more nephrotoxic in cases of preexisting renal functional disturbances and also that inner ear damage can occur more easily.

In cases of already existing conductive hearing disturbances, we could observe clinically and experimentally only an increase of the aminoglycoside antibiotic ototoxicity in the case of otitis media and to an even higher degree in the case of labyrinthitis.

Individual and Familial Sensitivity

An individual and familial sensitivity for ototoxic inner ear impairment exists no doubt. As to the data published recently and our personal clinical and experimental experience this individual and familial sensitivity, however, is not of that importance as was formerly presumed. In the experimental series in which we excluded nephrotoxicity by the administration of smaller doses, the big differences between the individual animals could be reduced. In spite of this, it is possible that among 10 animals treated with the same dose 9 have a slight hearing impairment and 1 animal a significantly higher damage. Because of the possibility of this individual sensitivity, no absolute guarantee can be given in clinical practice that no ototoxic damage will take place, even when treating with a classical aminoglycoside antibiotic and usual doses [24].

Pregnancy

A few cases of hearing disturbances in children have been described which probably have their origin in an administration of aminoglycoside antibiotics during pregnancy; however, many investigations do mention the slight ototoxic danger. Experimental investigations on the pharmacokinetics of aminoglycoside antibiotics in the fetus demonstrate that much lower aminoglycoside antibiotic concentrations penetrate into the fetal plasma and our histological studies in the guinea pig indicate that the

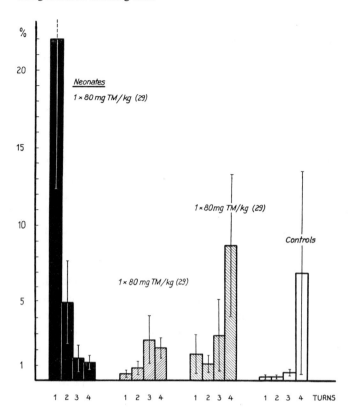

Fig. 5. Histocochleogram after 29 days of tobramycin (TM) in 1 group of 8 newborn guinea pigs and 2 groups of 4 adolescent and adult guinea pigs, respectively, as well as of one control group of 8 adult untreated guinea pigs.

inner ear damage is much higher in the dams than in the fetus [23]. Nevertheless, no increased ototoxicity could be found in the pregnant guinea pigs [23, 27]. During the first months of pregnancy, however, special attention is necessary because of teratologic reasons.

Newborns and Small Children

As to the ototoxic danger in the case of newborns and small children, the opinions vary. In experimental series in the guinea pig a statistically significant increase of the ototoxic risk of the aminoglycoside antibiotics in newborn animals could be found (fig. 5) [23]. This was also found in rats [47] and in cats [2].

Sound Exposure

The enhanced effect or at least the additive effect of the concomitant exposure to loud sounds and aminoglycoside antibiotics seems proved since the publication of *Darrouzet and De Lima Sobrinho* [9] and explained by the vasoconstriction of the spiral vessels of the basilar membrane and in this way the yet further limitations of energy supply of the cochlear hair cells [12, 38]. *Jauhiainen* et al. [34] demonstrated the interaction with sound and neomycin in guinea pigs and *Brown* et al. [4, 5] confirmed these observations and included kanamycin. *Hawkins* [29] made the interesting point that there was only an augmentation if the expected hearing lesions were located in the same range of frequencies. For the clinical situation we agree with *Brummett* [7] that it is wise to keep noise exposure of patients receiving the aminoglycoside antibiotics to a minimum.

Diuretics

Cases of temporary changes in hearing after a loop diuretic and especially ethacrynic acid and furosemide are well known since the end of the 1960s. There are also some clinical reports that ethacrynic acid and furosemide can produce permanent ototoxicity. These cases occurred mostly in patients who were also receiving an aminoglycoside antibiotic. These clinical suspicions have been verified in experimental animals. Whereas we could not demonstrate a potentiation of the ototoxicity of gentamicin and ethacrynic acid as well as furosemide if we used these drugs in dosages that were not ototoxic by themselves, *Brummett* et al. [6] and *Nakai* [39] demonstrated the interaction that takes place very rapidly after the administration of a single dose of ethacrynic acid to an animal pretreated with a single dose of kanamycin. The clinical data as well as the experimental data have developed the opinion that the potentiation of the ototoxicity of the two drugs may be expected if one of the two drugs is given in an ototoxic dosage, the diuretic is administered quickly, a high degree of renal functional disturbance is present and the interval between the application of the two drugs is less than 1–2 h [9, 18, 23, 30, 42].

Cephalosporins

Although the simultaneous administration of aminoglycoside antibiotics and cephalosporins leads to a decrease to the nephrotoxicity of the aminoglycoside antibiotics in the rat, a simultaneous therapy with a maximal dosage of aminoglycoside antibiotics and cephalosporins should only be performed when strictly indicated.

Treatment of Ototoxic Damage

A significant case of cochlear or vestibular ototoxicity in uremia will be treated as an emergency by means of hemodialysis, whereas in case of normal renal function dialysis is not of substantial help as demonstrate figures 3 and 4 on the serum and perilymph pharmacokinetics of dibekacin in the normal and bilaterally nephrectomized guinea pig. A therapeutic effect cannot surely be attributed to the low-dose heparine treatment. Experimental investigations have shown that hydro- and salidiuresis were not effective and dimercaptopropanol (BAL) is not effective in the treatment of ototoxic damage but has an intrinsic ototoxic potential [14, 24].

Cochlear and vestibular damage require the same treatment as a sudden hearing loss or vestibular disturbances. It consists mainly of rheo-macrodex, vitamin B complex and corticoids, Trental® or Dusodril®, whereas vertigo will be treated symptomatically but without a quinine derivative. Fortunately the present knowledge of the ototoxicity of aminoglycoside antibiotics makes ototoxic damage rare.

Summary

Today the number of potentially ototoxic substances is high, but the most important class is that of the aminoglycoside antibiotics. The clinical symptoms and signs of the aminoglycoside antibiotic ototoxicity are considered as well as the symmetry of these lesions, their possible reversability or progressiveness. The data collected up to now concerning the main factors influencing ototoxicity of the aminoglycoside antibiotics are perused. In special paragraphs the intrinsic ototoxic potential of the newer aminoglycoside antibiotics, the influence of the dosage and route of administration, of renal function, preexisting hearing disturbances, individual and familial sensitivity, pregnancy, newborn age and the combination with sound exposure, diuretics and cephalosporins are considered. Finally, the treatment of ototoxic damage is indicated.

References

1   Ballantyne, J.: Ototoxicity: a clinical review. Audiology *12:* 325–336 (1973).
2   Bernard, P.; Bourret, C.; Pechere, J.-C.; Remington, J.: Gentamicin inner ear susceptibility is higher in newborn than in adult cats. Curr. Chemother. Inf. Dis. *1:* 604–606 (1980).

3   Bodey, G. P.; Rodriguez, V.; Valdivieso, M.; Feld, R.: Amikacin for treatment
    of infections in patients with malignant diseases. J. infect. Dis. *134:* suppl., pp.
    421–427 (1976).
4   Brown, J. J.; Brummett, R. E.; Meile, M. B.; Vernon, J.: Combined effects of
    noise and neomycin. Acta oto-lar. *86:* 394–400 (1978).
5   Brown, J. J.; Brummett, R. E.; Fox, K. E.; Bendrick, T. W.: Combined effects
    of noise and kanamycin: cochlear pathology and pharmacology. Archs Otolar.
    (1980).
6   Brummett, R. E.; Brown, R. T.; Himes, D. L.: Quantitative relationship of the
    ototoxic interaction of kanamycin and ethacrynic acid. Archs Otolar. *105:* 240–
    246 (1979).
7   Brummett, R. E.: Drug-induced ototoxicity. Drugs *19:* 412–428 (1980).
8   Caussé, R.: Action toxique vestibulaire et cochléaire de la streptomycine au
    point de vue expérimental. Annls Oto-lar. *66:* 518–538 (1949).
9   Darrouzet, J.; De Lima Sobrinho, E.: Oreille interne, kanamycine et trauma-
    tisme acoustique: Etude expérimentale. Revue Lar. *83:* 781–806 (1979).
10 Darrouzet, J.; Guilhaume, A.: Ototoxicité cochléaire comparée de trois anti-
    biotiques: kanamycine, gentamicine, tobramycine. Etude histologique et ultra-
    structurale. Revue Lar. *97:* 655–673 (1976).
11 Falk, S. A.: Combined effects of noise and ototoxic drugs. Environ. Health
    Perspect. *2:* 5 (1972).
12 Federspil, P.: Über die klinische Ototoxicität des Gentamycins und ihre Rever-
    sibilität. Arch. Ohr.-Nas.-Kehlkheilk. *196:* 237–243 (1970).
13 Federspil, P.: Übersicht über die in Deutschland beobachteten Fälle von Gent-
    amycin-Ototoxizität. HNO *19:* 328–331 (1971).
14 Federspil, P.: Experimentelle Untersuchungen zur Ototoxizität des Gentamy-
    cins; Habilitationsschrift, Homburg (1973).
15 Federspil, P.: Pharmakokinetik des Gentamycins in Serum und Innenohrflüssig-
    keiten. Aktuelle Medizin. Münch. med. Wschr. *115:* 8 (1973).
16 Federspil, P.: Morphologische Untersuchungen zur Ototoxizität von Gentamy-
    cin und Tobramycin. ArzneimittelForsch. *23:* 1739 (1973).
17 Federspil, P.: Ototoxicität der Aminoglykosid-Antibiotica und Otitis media.
    Arch. klin. exp. Ohr.-Nas.-KehlkHeilk. *207:* 487–488 (1974).
18 Federspil, P.: Zur Ototoxizität der Aminoglykosid-Antibiotika. Referat vor
    der Paul-Ehrlich-Gesellschaft, 1. 10. 1976, Frankfurt. Infection *4:* 239–248
    (1976).
19 Federspiel, P.: Oto-Rhino-Laryngologie; in Kuemmerle, Garrett, Spitzy, Klini-
    sche Pharmakologie und Pharmakotherapie (Urban & Schwarzenberg, Mün-
    chen 1976).
20 Federspil, P.: Evaluation of the ototoxicity of netilmicin. Curr. Chemother.
    975: 976 (1978).
21 Federspil, P.: Ototoxicity; in The Royal Society of Medicine Round Table
    Discussion on Gentamicin and Tobramycin, pp. 15–27 (Academic Press, Lon-
    don/Grune & Stratton, New York 1978).
22 Federspil, P.; Schätzle, W.; Tiesler, E.: Pharmacokinetics and ototoxicity of
    gentamicin, tobramycin, and amikacin. J. infect. Dis. *134*: suppl., pp. 200–205
    (1976).

23 Federspil, P.: Die klinische Ototoxizität und ihre Prophylaxe; in Physiologi-
    sche und pharmakologische Grundlagen der Therapie, Berliner Seminar 2
    (Vieweg, Braunschweig 1978).

24 Federspil, P.: Antibiotikaschäden des Ohres (Barth, Leipzig 1979).

25 Federspil, P.; Schindler, K.; Weich, C.; Tiesler, E.; Schätzle, W.; Ziegler, M.:
    Zur klinischen Wirksamkeit, Ototoxizität und Nephrotoxizität des Amikacins.
    Infection 7: 81–87 (1979).

26 Federspil, P.: Das Ohr als Reaktionsorgan bei chronischer Medikation. AMI-
    Berichte 1: 165–168 (1980).

27 Federspil, P. J.; Schätzle, W.; Kayser, M.; Sack, K.; Schentag, J.: Influence of
    total dose, division of daily dose, age, and pregnancy on aminoglycoside oto-
    toxicity. Curr. Chemother. Inf. Dis. 1: 607–108 (1980).

28 Federspil, P.: Experimentelle Untersuchungen zur Ototoxizität der Aminogly-
    kosid-Antibiotika und ihre klinische Bedeutung. Laryng.-Rhinol. 60: 587 (1981).

29 Hawkins, J. E., Jr.: Comparative otopathology: aging, noise, and ototoxic drugs.
    Adv. Oto-Rhino-Laryng., vol. 20, pp. 125–141 (Karger, Basel 1973).

30 Hawkins, J. E., Jr.: Drug ototoxicity. Handbook of sensory physiology, vol.
    v/3 (Springer, Heidelberg 1976).

31 Ilberg, C. von: Toxische Schäden des Hörorgans; in Berendes, Link, Zöllner,
    Hals-Nasen-Ohren-Heilkunde, vol. 6, pp. 43.1–24 (Thieme, Stuttgart 1980).

32 Jackson, G. G.; Arcieri, G. M.: Ototoxicity of gentamicin in man: a survey
    and controlled analysis of clinical experience in the United States. J. infect.
    Dis. 124: suppl., pp. 130–137 (1971).

33 Jackson, G. G.: Present status of aminoglycoside antibiotics and their safe, ef-
    fective use. Clin. Ther. 1: 371–387 (1977).

34 Jauhiainen, J.; Kohonen, A.; Jauhiainen, M.: Combined effect of noise and
    neomycin on the cochlea. Acta oto-lar. 73: 387–390 (1972).

35 Höffler, D.: Klinik und Pharmakologie des Gentamycins. Berl. Ärztekammer
    1: 13 (1969).

36 Lehnhardt, E.: Zur Ototoxizität der Antibiotica. HNO 18: 97–101 (1970).

37 Meyers, R. M.: Ototoxic effects of gentamicin. Archs Otolar. 92: 160–162
    (1970).

38 Misrahy, G. A.; Spradley, J. F.; Dzinovic, S.; Brooks, C. J.: Effect of intense
    sound, hypoxia and kanamycin on the permeability of cochlear partitions. Ann.
    Otol. 70: 572 (1961).

39 Nakai, Y.: Combined effect of 3',4'-dideoxykanamycin B and potent diuretics
    on the cochlea. Laryngoscope, St Louis 87: 1548–1558 (1977).

40 Oldershausen, H.-F. von; Ullmann, U.; Dürr, R.; Zysno, E.; Caesar, F.; Au-
    wärter, W.: Clinical and bacteriological studies on short- and long-term treat-
    ment with high doses of gentamycin. Progress in Antimicrob. and Anticancer
    Chemother. Proc. 16th Int. Congr. of Chemother., Tokyo 1970.

41 Pfaltz, C. R.; Herzog, H.; Staub, H.; Way, W.: Zur ototoxischen Wirkung ho-
    her Streptomycindosen. Schweiz. med. Wschr. 51: 1472–1480 (1960).

42 Prazma, J.; Browder, J. P.; Fischer, N. D.: Ethacrynic acid ototoxicity poten-
    tiation by kanamycin. Ann. Otol. 83: 111–118 (1974).

43 Strauss, P.; Rosin, H.; Quante, M.; Harari, G.; Winter, V.; Löbner, S.: Gent-
    amycin im Verteilungsgleichgewicht zwischen Serum, Perilymphe und Liquor

beim Meerschweinchen nach Dosierung im therapeutischen Bereich. Archs Oto-Rhino-Lar. *218:* 79–86 (1977).

44 Sweedler D. R.; Gravenkemper, C. F.; Bulger, R. J.; Brodie, J. L.; Kirby, W. M. M.: Laboratory and clinical studies on gentamicin. Antimicrob. Agents Chemother. *1:* 157 (1963).

45 Thompson, W. L.: Gentamicin and tobramycin nephtrotoxicity in dogs given continuous or once-daily intravenous injections. Assessment of aminoglycoside toxicity, p. 23, Bürgenstock 1977.

46 Tompsett, R.: Relation of dosage to streptomycin toxicity. Ann. Otol. *57:* 181 (1948).

47 Uziel, A.; Romand, R.; Gabrion, J.: Intrauterine ototoxicity of kanamycin in the guinea pig. Inserum *68:* 347–358 (1977).

48 Wersäll, J.: The ototoxic potential of netilmicin compared with amikacin. Scand. J. infect. Dis. *23:* suppl., pp. 104–113 (1980).

Prof. Dr. med. P. Federspil, Hals-Nasen-Ohren-Universitätsklinik,
D-6650 Homburg/Saar (FRG)

Adv. Oto-Rhino-Laryng., vol. 27, pp. 159–167 (Karger, Basel 1981)

Sudden Unilateral Loss of Vestibular Function

C. R. Pfaltz, A. Meran

Department of Oto-Rhino-Laryngology, University Hospital, Basel, Switzerland

Apart from the classic attack of Menière's disease, which may be regarded as a well-established nosologic entity of cochleovestibular symptoms, cochlear or vestibular disorders of sudden onset but of uncertain origin may occur independently from each other. The symptomatology of the vestibular disturbances is particularly dramatic and alarms both the patient and his physician who is consulted in an emergency situation. It is not uncommon that the patient is transferred to the nearest hospital labelled with the diagnosis of acute intoxication, stroke or cardiac infarction.

Vertiginous attacks without cochlear symptoms are very often designated as 'Menière's syndrome', a completely erroneous classification which is only used to hide medical ignorance. Not so long ago a new term was introduced to classify these acute vestibular disorders more correctly: *vestibular neuronitis*. Using this very specific term we must, however, bear in mind the original definition of this particular vestibular disturbance, made by *Dix and Hallpike* [1952]: 'Some form of organic disease confined to the vestibular apparatus and localized in all probability to its peripheral nervous pathways up to and including the vestibular nuclei in the brainstem.'

According to their original definition and the diagnostic label chosen by these authors it must be assumed that the underlying pathogenetic mechanism of the 'organic disease confined to the vestibular apparatus' is of inflammatory origin. Since in most cases substantial evidence is lacking, the term vestibular *'neuronitis'* has been criticized and more general terms such as 'sudden loss of vestibular function' [*Lindsay and Hemenway,* 1956] or 'vestibular paralysis of sudden onset' [*Hart,* 1965] were introduced. In order to investigate the problem of etiology, pathogenesis

Table I. General symptomatology of sudden loss of vestibular function

---

Hearing normal
Tinnitus absent
Vertigo: sudden, dramatic attack, combined with nausea and vomiting; followed by
   a state of permanent disequilibrium, gradual but slow recovery
Consciousness not impaired
Visual disturbances absent
Neurologic symptoms absent

Table II. Vestibular symptomatology

Spontaneous nystagmus → normal side
Postural changes enhance spontaneous nystagmus
Vestibulospinal reflexes pathologic on the side of the lesion
Caloric test: responses abolished on the side of the lesion
Galvanic test: in end organ lesions normal and symmetric responses – in retrolaby-
   rinthine lesions pathologic responses
Optokinetic test: in periphereal lesions normal responses – in central lesions patho-
   logical responses

---

and localization of the vestibular lesion more thoroughly, we have examined and followed up 100 patients showing initially the following typical
vestibular symptomatology (table I, II).

Vestibular Symptomatology

At first sight this vestibular syndrome seems to be rather uniform but
the question arises whether it may be considered as a *nosologic entity*.
This hypothesis, however, can only be accepted if the functional disorder
is: located without exception to the same structures of the vestibular system; caused by the same pathogenetic factor; taking the same course, and
showing the same age and sex distribution.

Distribution of Age and Sex. The highest incidence of morbidity is
found between 30 and 40 years in the female group and between 50 and
60 years in the male group. There is no evidence of a statistically significant difference with respect to sex distribution.

Localization of the Vestibular Lesion. Caloric responses indicate the side of an unilateral vestibular disorder by means of a unilateral decrease of abolition of vestibular responses, corresponding with the impaired ear. The caloric test results, however, do not reveal the site of the lesion whereas *the galvanic test [Pfaltz, 1969]* may clearly indicate whether the functional disorder is located within the end organ, the peripheral neuron or within the central vestibular pathways.

End organ lesion
Caloric responses: unilaterally reduced or abolished
Galvanic responses: normal and symmetric

Retrolabyrinthine lesion (peripheral type):
Caloric responses: unilaterally abolished
Galvanic responses: ipsilaterally impaired
                        (threshold difference between left and right responses > 4 mA – sometimes completely abolished)

Retrolabyrinthine lesion (central type):
Caloric responses: unilaterally abolished
                        (eventually inhibition or total suppression of vestibular nystagmus elicited from the normal labyrinth, but directed towards the side of the lesion)

Galvanic responses: either raised threshold or abolition of responses induced on the side of the lesion; or complete abolition of galvanic nystagmus directed towards the side of the lesion (induced by unilateral stimulation on both sides)

Results
Labyrinthine lesions   13 %
Retrolabyrinthine lesions (peripheral type)   38 %
Retrolabyrinthine lesions (central type)   49 %

Pathogenetic Factors. The most striking observation made in our series of 100 cases of unilateral sudden loss of vestibular function is the great incidence of active infections (64 %, table III); viral and microbial infections of the upper respiratory tract are prevailing. We have been able to corroborate our clinical findings by the following laboratory test results:

Pathologic sedimentation rate   40 % +
Specific serologic tests (microbial and viral infections)   43 % +
Pathologic serum protein electrophoresis   36 % +
Pathologic CSF   5/17 cases
Positive dye test   1/17 cases

Table III. Pathogenetic factors: infections

	Cases (%)
Sinusitis (purulent)	25 (40)
Influenza	15 (24)
Tonsillitis	7 (10)
Dental infection (granuloma)	4 (7)
Bronchitis	5 (8)
Gastroenteritis	6 (9)
Parotitis	1 (1)
Toxoplasmosis	1 (1)

Table IV. Pathogenetic factors: vascular damage

	%
Severe chronic hypertension	14
Myovascular insufficiency	9
Vertebrobasilar insufficiency	3
Diabetes	10
Total	36

The second, most important factor which might be involved in the pathogenesis of unilateral sudden loss of vestibular function is the *impairment of blood supply* either at the level of the end organ, the peripheral nerve or the central vestibular pathways in the brainstem. 26% of our patients show definite symptoms of vascular damage alone and 10% in combination with a severe diabetic angiopathy (table IV). Other factors which might be correlated with the pathogenesis of this particular vestibular disorder are: autoimmune disease (3 cases) and drug incompatibility reactions (9 cases).

Synopsis of Clinical and Laboratory Findings. On the basis of a topic diagnosis of the unilateral vestibular lesion and the course of the symptomatology 4 groups are distinguished (table V–VIII).

Table V. Labyrinthine (end organ) lesion: 13 %

Group A
Caloric test: pathologic responses
Galvanic test: normal responses
Pathogenetic factors uncertain
Spontaneous functional restitution
   complete and rapid

Table VI. Retrolabyrinthine lesion: 23 %

Group B (peripheral type – partial loss of vestibular function)
Caloric test: pathologic responses
Galvanic test: pathologic responses
Pathogenetic factors: 6/23 infections
                         5/23 diabetes
Spontaneous functional restitution
   complete, but delayed

Table VII. Retrolabyrinthine lesion: 15 %

Group C (peripheral type – total loss of vestibular function)
Caloric test: pathologic responses
Galvanic test: pathologic responses
Pathogenetic factors: 15/15 infections
                         3/15 diabetes
Spontaneous functional restitution
   delayed and incomplete (3/15)

Table VIII. Retrolabyrinthine lesion: 49 %

Group D (central type – unidirectional abolition of caloric and galvanic nystagmus)
Caloric and galvanic test responses pathologic
Optokinetic responses: sometimes pathologic
Pathogenetic factors: 32/49 infections
                             23/49 vascular damage
Spontaneous functional restitution
   Delayed: 29/49
   Incomplete or absent: 20/49

Discussion

Unilateral sudden loss of vestibular function is a *vestibular syndrome* without cochlear accompaniments, which is due to a lesion either of the vestibular end organ, the vestibular nerve or the central vestibular pathways.

This vestibular syndrome may not be considered as a nosologic entity because its underlying pathogenetic mechanisms seem either uncertain or inhomogenous and the site of the lesion is not constant. For those reasons terms such as 'vestibular neuronitis' are misleading because 'neuronitis' implies an inflammatory process involving only one part of the vestibular system, i. e. the first neuron. Only some of the cases belonging to group B and C (table VI, VII) demonstrating the characteristic clinical and laboratory findings which indicate a major pathogenetic role of an infection (e. g. toxoplasmosis) or of an immunologic disorder, may be classified as 'vestibular neuronitis sensu strictiori'. It is rather difficult to reduce the various pathogenetic factors (infection, immunologic disorders, vascular damage, drug incompatibility) to one *pathogenetic system*.

Viral Infection. According to the experimental studies of *Temesrekasi* [1970] and *Temesrekasi and Alram* [1969] and those of *Hoyle* [1954] the virus of epidemic parotitis and of influenza is known to cause lesions of the endothelial layer, particularly in smaller vessels such as arterioles, pre- and postcapillaries. Even temporary or partial occlusion of these small vessels will result in microcirculatory disorders, either due to *sludging* (reversible) or to *microembolism* (irreversible). This may occur in the end organ, in the peripheral nerve and in the brainstem as well. (Moreover, this type of virus is neurotropic and may even produce a meningo-encephalitis or a serous labyrinthitis with secondary degeneration of the afferent nerve fibers).

Autoimmune diseases cause granulomatous tissue reactions and particularly vasculitis. Recently *McCabe* [1979] has published a paper on autoimmune sensorineural hearing loss. He points out that in these cases vestibular function is involved as well and that tissues beyond the inner ear may become affected, e. g. the facial nerve. The underlying pathogenetic mechanism is a chronic vasculitis, which has been confirmed histologically. Hence it may be assumed that immunologic disorders may result in microcirculatory disturbances due to chronic microvasculitis, involving not only the vestibular end organ but in some cases also the peripheral nerve and the central pathways in the brainstem.

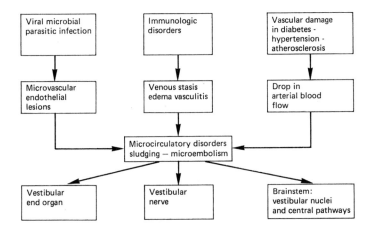

Fig. 1. Model of interaction of multiple pathogenetic factors in cases of unilateral sudden loss of vestibular function.

Microcirculatory disorders in the brainstem, due to vertebrobasilar insufficiency, may be one of the major pathogenetic factors responsible for *sudden loss of central vestibular function.* A transient decrease in cerebral blood flow without actual brainstem infarction is known as *insufficiency.* In this state, cerebral blood flow is unable to satisfy the metabolic requirements of the brain for a brief period of time. Symptoms and *functional deficits,* suffered by the patient during that period, *are reversible* and no permanent pathological changes or neurological sequelae are sustained. A broader range of mechanisms may be considered as etiologic: hypotension, cardiac arrhythmia, decreased cardiac output, microembolism, altered blood viscosity and stenosis of the extracranial cerebral arteries. Both atherosclerosis and thrombosis in extracranial arteries may cause symptoms either by mechanical occlusion or by acting as sources for embolization of platelet or cholesterol material [*Burns,* 1973]. Hemodynamically the result of a drop in blood flow through the vessels supplying the brainstem is a reduced flow in the anterior and posterior inferior cerebellar arteries, which supply the lateral tegmentum of the brainstem, including the vestibular nuclei. Their status as end arteries and the steep pressure gradient from the aorta have been used to explain the vulnerability of this particular brainstem area if systemic pressure fails [*Burns,* 1973]. We have been able to provide substantial evidence of a severe vascular damage in 36 % of our 100 cases and in 46 % of group D (table VIII; central

vestibular lesions); hence we may assume that microcirculatory disturbances have to be taken into consideration as an original cause of a sudden unilateral loss of vestibular function in our series of patients. Most
probably this intriguing vestibular syndrome is not caused by one single
pathogenetic factor but by the interaction of several pathologic mechanisms, as suggested hypothetically in figure 1.

Treatment

In cases of proven autoimmunologic disorders, long-term administration of corticosterone is the method of choice. In cases suggesting a microcirculatory disorder, repeated intraveneous infusion of low molecular dextran are recommended. Vestibular rehabilitation after unilateral loss of
vestibular function is accelerated by physical exercises and optokinetic
and/or vestibular training [*Pfaltz,* 1977].

Prognosis depends primarily on the origin and the site of the lesion,
in the second place on the interval between the onset of the symptoms and
the beginning of the treatment, and furthermore on the age of the patient.
Senile involution delays spontaneous recovery and deteriorates prognosis
considerably [*Meran and Pfaltz,* 1979]. Recurrences are extremely rare.

Summary and Conclusions

In a study of 100 patients, showing the typical symptomatology of a
unilateral sudden loss of vestibular function, who have been thoroughly
examined and controlled over a longer period of time (1969–1979), the
authors have made the following observations: Symptomatology in unilateral loss of vestibular function is predominantly uniform and characteristic. The vestibular findings and particularly the different course of central vestibular compensation indicate that there is no uniform localization
of the vestibular lesion and no conformity with respect to the presumable
pathogenetic mechanisms. Moreover, etiology and pathogenesis are still
rather uncertain. Vascular pathology, infections and autoimmune diseases
must be taken into account as primary triggering factors.

The authors presume that in many cases microcirculatory disorders
are the primary cause of this particular vestibular disorder. For this reason, the term 'vestibular neuronitis' is considered to be misleading, be-

cause only in a few cases enough substantial evidence can be provided indicating the primary pathogenetic role of inflammation or infection in a retrolabyrinthine, neural vestibular lesion. Hence, it may be concluded that *unilateral sudden loss of vestibular function may not be considered as a nosologic entity but must be interpreted as a characteristic vestibular syndrome,* caused by the interaction of various pathologic mechanisms acting on different parts of the vestibular system.

References

Burns, R. A.: Basilar vertebral artery insufficiency as a cause of vertigo; in Wolfsson, Otolaryng. Clinics of North America Symp. on Vertigo, pp. 287–300 (Saunders, Philadelphia 1973).

Dix, M. R.; Hallpike, C. S.: The pathology symptomatology and diagnosis of certain common disorders of the vestibular system. Ann. Otol. Rhinol. Lar. *61:* 987 (1952).

Hart, C. W. J.: Vestibular paralysis of sudden onset and probably viral etiology. Ann. Otol. Rhinol. Lar. *74:* 33–47 (1965).

Hoyle, L.: The release of influenza. Virus from the infected cell. J. Hyg., Lond. *52:* 180 (1954).

Lindsay, J. R.; Hemenway, W. C.: Postural vertigo due to unilateral sudden partial loss of vestibular function. Ann. Otol. *65:* 692–707 (1956).

McCabe, B. F.: Autoimmune sensorineural hearing loss. Ann. Otol. Rhinol. Lar. *88:* 585–589 (1979).

Meran, A.; Pfaltz, C. R.: Der akute Vestibularisausfall. Akt. neurol. *6:* 27–38 (1979).

Pfaltz, C. R.: The diagnostic importance of the galvanic test in otoneurology. Pract. oto-rhino-laryn. *31:* 193–203 (1969).

Pfaltz, C. R.: Vestibular habituation and central compensation. Adv. Oto-Rhino-Laryng., vol. 22, pp. 136–142 (Karger, Basel 1977).

Temereskasi, D.: Zur Pathogenese des akut entstehenden isolierten Vestibularisausfalles. HNO *18:* 313–316 (1970).

Temesrekasi, D.: Zur Pathogenese des akut entstehenden isolierten Vestibularisaustitis epidemica. Mschr. Ohrenheilk. Lar.-Rhinol. *103:* 366–375 (1969).

Prof. Dr. C. R. Pfaltz, Head of the Department of Oto-Rhino-Laryngology, University Hospital, CH-4031 Basel (Switzerland)

Adv. Oto-Rhino-Laryng., vol. 27, pp. 168–175 (Karger, Basel 1981)

Provoked Vestibular Nystagmus and Caloric Reactions after Sudden Loss of Vestibular Function

H. Scherer

ENT Clinic of the University of Munich, Munich, FRG

After a sudden loss of vestibular function patients usually show clear clinical signs and subjective symptoms.

Spontaneous and Provoked Nystagmus in the Early Days

In the early days we find: (a) A spontaneous nystagmus (paralytic nystagmus) to the contralateral healthy side, which – when graded with Alexander's classification is of second – or in the first days – of third degree. (b) A direction persistent, provoked, usually horizontal nystagmus. Depending upon the local damage it can be combined with a rotary component. This is probably due to a lesion of more than one semicircular canal [4] and to functional connections between the canal and the otolitic system [5]. A reversal of provoked nystagmus or a vertical nystagmus never occurs in this stage [15], (c) A facilitation of spontaneous nystagmus when the patient lies on the affected side. We find an inhibition when the patient lies on the healthy side. Patients with severe subjective symptoms therefore usually suffer less when lying on the contralateral side of the lesion [5].

Spontaneous and Provoked Nystagmus in Later Days

When the damage is exclusively peripheral, vestibular compensation starts with short or no delay. This compensation reduces the spontaneous nystagmus exponentially by means of efferent vestibular fibers [8, 10, 12,

Table I. Grades of compensation after a sudden loss of vestibular function

Grade	Definition	Signs and symptoms
IV	the compensation is complete	*no signs* – except loss of function in caloric test; *no symptoms*
III	the compensation is virtually complete	*head-shaking nystagmus;* direction preponderance in caloric test; unsteadiness during fast head movements
II	the compensation is fairly complete	*spontaneous nystagmus in ENG,* eyes open in total darkness; provoked nystagmus with Frenzel's glasses; tendency to fall or deviate in difficult types of vestibulospinal tests; unsteadiness when doing unaccustomed movements
I	the compensation is incomplete	*spontaneous nystagmus with Frenzel's glasses;* marked direction preponderance in experimental vestibular tests; provocative, direction-constant nystagmus; tendency to fall or deviate in vestibulospinal tests
0	no compensation	*no decrease* of spontaneous nystagmus; *no decrease* in the tendency to fall

13]. This compensatory mechanism is probably localized in the vestibular nuclei [7, 11, 12] and possibly in other regions, i. e. the nucleus motorius tegmenti as shown in animal experiments [17].

The decline of the intensity of spontaneous nystagmus, its speed and the time between onset and offset of the nystagmus cannot be evaluated statistically in humans, because on the one hand usually nystagmus-reducing drugs are prescribed in the first weeks and sometimes months. On the other hand, particularly in the case of a vascular disease, we cannot definitely exclude an accompanying central defect, which would hinder the compensation [6, 14] as well as an otolitic defect. In animal experiments, however, those problems do not exist.

Independent from central vestibular compensation there are other mechanisms, which try to improve the person's equilibrium. The most important ones are working via the optokinetic system and the somatosensory system. For the compensation of a sudden loss of vestibular function we therefore require not only an undamaged central vestibular system but in addition much larger regions of brain stem, of the cerebellum and of optokinetic pathways in mesencephalon and cerebrum should remain undamaged [12, 13]. In elderly patients with cerebral sclerosis we often find a reduced compensation because of the likelihood of these areas being involved in the cerebral degeneration.

The amount of compensation in case of a stationary loss of function can be graded depending upon the clinical signs and symptoms (table I).

There is No Compensation (Grade 0). When the intensity of the spontaneous nystagmus after a sudden loss of vestibular function does not show any decrease; when the tendency to fall or to deviate to the affected side in vestibulospinal tests does not decrease.

The Compensation is Incomplete (Grade I). As long as we see a spontaneous nystagmus with Frenzel's glasses. Usually this spontaneous nystagmus causes a marked directional preponderance in all tests. Young persons may show good results in vestibulospinal tests due to their effective somatosensory system.

The Compensation is Fairly Complete (Grade II). When we *cannot* see a spontaneous nystagmus with Frenzel's glasses; but we can observe a direction constant provoked nystagmus with Frenzel's glasses to the healthy side; and we see a spontaneous nystagmus in the electronystagmogram in total darkness with open eyes.

We also see a directional preponderance not only in unphysiological tests (i. e. caloric test) but also in physiological tests (i. e. rotatoric or pendular tests). The patients, most often elderly, reveal a tendency to fall or deviate to the affected side in *difficult* types of vestibulospinal tests. Patients regularly feel an unsteadiness when carrying out unaccustomed movements.

The Compensation is Virtually Complete (Grade III). When we cannot find a spontaneous nystagmus with Frenzel's glasses. When we cannot find a spontaneous nystagmus in electronystagmogram in total darkness with open eyes. When we find an equal or nearly equal response during physiological experimental tests (pendular or rotatoric tests). When we cannot observe a tendency to fall or deviate in vestibulospinal tests.

But we still see more than one sign of the labyrinthine defect: (a) a loss of function or a marked sidedifference of excitability in the caloric

tests; (b) a directional preponderance to the unaffected side in unphysiological tests (caloric test); (c) a head-shaking nystagmus; (d) an unsteadiness during fast head movements.

The Compensation is Complete (Grade IV). When we cannot find a
spontaneous nystagmus with Frenzel's glasses. When we cannot find the
spontaneous nystagmus in electronystagmogram anymore. When we cannot find a difference of sides in physiological experimental tests. When we
cannot find a tendency to fall or deviate to the affected side even in difficult types of vestibulospinal tests. When we cannot find any provocative
nystagmus.

A total loss of vestibular function will usually cause a compensation
of up to grade III. A partial loss of vestibular function can reach grade IV.

Another grading of vestibular compensation according to *different
kinds* of compensatory mechanisms was published by *Pfaltz* et al. [10].
(1) The *stage of deficiency* (spontaneous nystagmus and very marked DP
of galvanic nystagmus towards the side of the normal labyrinth). (2) *The
stage of central compensation* (absence of vertigo, disequilibrium and
spontaneous or positional nystagmus, symmetrical galvanic reactions near
and above threshold). (3) The *stage of recovery* (absence of vertigo, but
spontaneous and positional nystagmus as well as DP of galvanic nystagmus
directed towards the side of the lesion).

It is a rule that: there is no correlation between the intensity of the
labyrinthine lesion and the rate of compensation [7, 14]; there is a negative
correlation between age and speed of compensation; there is a positive
correlation between body activity and speed of compensation [3, 16]; high
doses of alcohol and sometimes emotion can cause decompensation.

The Role of the Caloric Test

The caloric test is the basic diagnostic tool in measuring a sudden loss
of vestibular function. But its role should not be overestimated as long as
we see a spontaneous nystagmus. During this time it is nearly impossible
to say whether a patient is suffering from a total or partial loss of function
[1]. A residual function may be undetected, because on the one hand, a
warm stimulus of 44 °C water temperature, which could invert the spontaneous nystagmus from the affected side, cannot be increased. On the
other hand, the acceleration of a spontaneous nystagmus by a cold stimulus

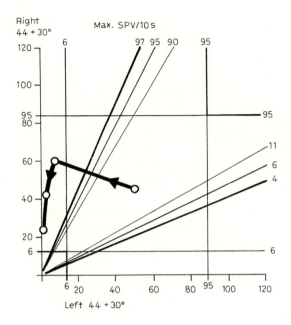

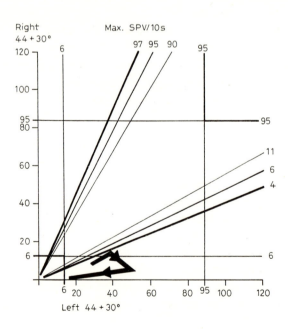

from the affected side can be of unspecific origin like any other provocative manoeuvre. The addition and subtraction of the spontaneous nystagmus to and from the results of the caloric test [14] brings better but still insufficient results.

At first (fig. 1, 2) we find an apparent increased reaction of the healthy labyrinth, because the spontaneous nystagmus is facilitated by the warm stimulus of this side. The increasing central compensation then causes a gradual reduction of the healthy side's excitability, called 'compensatory hypoexcitability'. This leads to a reduction of the difference between both sides and so to a reduction of spontaneous nystagmus. This development was confirmed by *Wolfe and Kos* [18] experimentally in rhesus monkeys and statistically by *van den Calseyde* et al. [2] in man. Inter- and intraindividual varieties, hidden central defects and other influences can sometimes cause different and apparent unphysiological processes.

The Role of Physiological Vestibular Tests
(Pendular, Rotatory and Others)

The physiological tests, especially the pendular test, vary less than the caloric test. They are particularly suitable for measuring the compensation. Figure 3 shows the result of a pendular test with increasing and decreasing stimuli. During slow acceleration we often see a directional preponderance to the right, which disappears during high stimuli. This directional preponderance during slow physiological accelerations shows us that the patient should still be localized in grade II (compensation is fairly complete) of vestibular compensation. High stimuli alone therefore are not able to demonstrate this sign.

Fig. 1. The idealized course of the healthy side's excitability (thick line). This line is inserted in a drawing of the ratio of responses from the right and left labyrinths in normals [9]. The numbered lines mark the interquantil areas (i. e. between line 90 and 11 we find the ratios of 80 % of normals).

Fig. 2. The course of ratio of responses from the right and left labyrinth of a patient with unilateral loss of vestibular function. Time interval between each caloric test: 4 weeks.

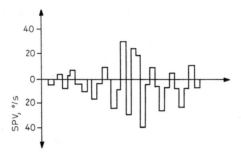

Fig. 3. Slow phase velocity of nystagmus during a pendular test with increasing and decreasing stimulation intensity. The patient, 8 weeks after a sudden loss of vestibular function, is in grade II of vestibular compensation.

Conclusion

In early days of a sudden loss of vestibular function the diagnosis can be easily obtained by observing the spontaneous nystagmus and by observing the reduction of labyrinthine excitability in the caloric test. The determination of absolute excitability however can be difficult. The caloric test's variability, nystagmus-reducing drugs or accompanying central vestibular disorders may masque the true situation.

In grading the central vestibular compensation we should therefore use more than one clinical test, especially provocative tests, vestibulospinal tests and physiological stimulations of the labyrinths in addition to the basic caloric stimulation.

References

1   Bönninghaus, H. G.; Frank, M.: Nystagmusuntersuchungen bei Pendelreizung nach einseitigen Labyrinthausfällen. Lar. Rhinol. Otol. *10:* 623 (1970).
2   Calseyde, P. van de; Ampe, W.; Depondt, M.: Use of the torsion swing test in a follow-up study of the ENG-registration in cases of unilateral labyrinthectomie. ORL *35:* 83 (1973).
3   Cawthorne, T.: Vestibular injuries. Proc. R. Soc. Med. *39:* 270 (1946).
4   Decher, H.; Schulte, H. G. Nachweis isolierter Bogengangsläsionen durch Drehreizschwellenteste. Arch. klin. exp. Ohr.-Nas.-KehlkHeilk. *195:* 1 (1969).
5   Fluur, E.: The otolith organs and the nystagmus problem (I–IV). Interaction between the utricles and the horizontal semicircular canals. Acta oto-lar. *75* (1973).

6   Ganz, H.: Plötzliche Ausfälle der Labyrinthfunktion. HNO 9: 89 (1961).
7   Kornhuber, H.: Physiologie und Klinik des zentral-vestibulären Systems. Arch.
     klin. exp. Ohr.-Nas.-KehlkHeilk. 194: 111 (1969).
8   Mittermaier, R. Die experimentellen Gleichgewichtsprüfungen; in Berendes,
     Link, Zöllner, Handbuch der HNO-Heilk., vol. III/3 (Thieme, Stuttgart 1965).
9   Mulch, G.; Scherer, H.: Methoden zur Untersuchung des vestibulären Systems.
     II. Thermische Prüfung. HNO-Informationen 5: 7 (1980).
10   Pfaltz, C. R.; Piffko, P.; Mishra, S.: Central compensation of vestibular dys-
     function. ORL 35: 71 (1973).
11   Pfaltz, C. R.; Piffko, P.: Central compensation of retrolabyrinthine lesion. Acta
     oto-lar. 73: 183 (1972).
12   Pfaltz, C. R.: Vestibular habituation and central compensation. ORL 22: 136
     (1977).
13   Pfaltz, C. R.; Kamath, R.: The problem of central compensation of peripheral
     vestibular dysfunction. Acta oto-lar. 71: 266 (1971).
14   Reker, U.: Akute isolierte Vestibularisstörungen. HNO 25: 122 (1977).
15   Stenger, H. H. Vestibularisstörungen in der Praxis; in Berendes, Link, Zöllner,
     Handbuch der HNO-Heilk., vol. V (Thieme, Stuttgart 1975).
16   Stierle, J. L.; Conraux, C.: La rééducation du vertigineux. Méd. Hyg., Genève
     35: 3392 (1977).
17   Strutz, J.; Schmidt, C. L.; Stürmer, C.: Origin of efferent fibers of the vesti-
     bular apparatus in goldfish. A horseradish peroxidase study. Newsci. Lett. 18: 5
     (1980).
18   Wolfe, J. W.; Kos, C. M.: Nystagmic responses of the rhesus monkey to ro-
     tational stimulation following unilateral labyrinthectomy. Trans. Am. Acad.
     Ophthal. Oto-lar. 84: ORL-38–ORL-45 (1977).

H. Scherer, MD, ENT Clinic of the University of Munich,
Klinikum Grosshadern, D-8000 Munich 70 (FRG)

Adv. Oto-Rhino-Laryng., vol. 27, pp. 176–189 (Karger, Basel 1981)

Positional Test in Acute Vestibular Disorders

T. Haid

ENT Clinic at the University of Erlangen-Nuremberg, Erlangen, FRG

Introduction

In the vestibular diagnosis, the positional test with quantitative assessment of the results makes an important contribution to the appraisal and identification of the severity of a disease and to its prognosis. Three acute diseases, i. e. vestibular neuronitis, Menière's disease and sudden hearing loss with vestibular involvement have been selected to demonstrate the usage of the positional test.

Material and Method

The results of the positional test from the average year 1978 were evaluated. From October 1973 to December 1979, a vestibular function test was carried out in 266 patients affected by a sudden hearing loss. In 1978, a total of 43 patients with this diagnosis was referred for examination. From October 1973 up to, and including, December 1979, the diagnosis of vestibular neuronitis was established in 154 persons. In 1978, this finding was made in 30 subjects. Among our patients, 210 presented with Menière's disease during the same period and, in 1978, the same disease was identified in 32 persons. After establishing the patient's history and examination of the spontaneous, gaze and head-shaking nystagmus, the positional test was carried out prior to the caloric test: At first, the patient was brought into the *head-hanging position;* subsequently, the re-

Table I. Maximum nystagmic intensity during the positional test to which 30 patients with vestibular neuronitis were subjected

Positional nystagmus: nystagmus beat during 60 s	Path./number 7/30	Positioning nystagmus: nystagmus beat during (s):		Path./number 16/30
20	0	5	5	2
40	2	5	10	4
60	0	10	10	4
80	1	10	15	1
100	0	10	20	2
120	0	15	10	0
140–160	4	15	15	1
		15	20	0
		20	15	0
		20	20	1
		20	25	0
		25	30	1

Patients with nystagmus in the positional test: 23/30 (77 %).

cumbent patient was told to *turn his head* at first to the right and then to the left; thereafter, he had to *turn his body* to the right, then to the left and finally he had to rise fast into a sitting posture. These positions were repeated three times. This allowed the examiner to determine firstly the reproducibility of the nystagmus and secondly to detect nystagmus reliably as this disorder sometimes was not identified until the second or third test. The investigation was always performed with the help of the Frenzel spectacles [*Kornhuber*, 1966], but also with ENG recording. In order to keep within the scope of this paper, only the positional test results of the three diseases will be discussed.

Results

Patients affected by vestibular neuronitis produced most frequently a pathological result during the positional test (77 %), followed by subjects afflicted with sudden hearing loss (35 %) and by those suffering from Menière's disease (31 %). Vestibular neuronitis and the sudden hearing loss revealed mostly a direction-fixed nystagmus during the positional test, while in the case of Menière's disease the direction-changing nystagmus accounted for 50 %. During the positional test, the nystagmus was in many cases beating toward the healthy ear when vestibular neuronitis and sudden hearing loss were present. By contrast, with Menière's disease, during the positional test, a direction-changing nystagmus occurred or a form beating toward the diseased ear (a particularly striking finding during the stage of irritation). A vertical nystagmus, which was observed four times with sudden hearing loss, may also point to a central reaction.

Common to all three diseases was that both the positional and positioning nystagmus were identified relatively frequently by positioning the patient on the affected and also on the intact side. During the stage of irritation of Menière's disease, patients often produced a nystagmus, a relatively intense positional form, only when lying on the diseased side (fig. 12).

With all three diseases, the positional nystagmus or the positioning type was identified with few exceptions, in at least two or more positions. The number of positions in which nystagmus developed depended on the stage of disease and also on the date of the vestibular function test. The maximum value, i. e. the maximum intensity of the positional nystagmus or positioning nystagmus was most pronounced with vestibular neuronitis. During the acute and subacute stage, 4 patients (table I) gave even a beat rate from 140 to 160/min for the positional nystagmus (average value: 100/min). Second place was held by the stage of irritation in 3 patients with Menière's disease giving an average value of 75 beats/min (maximum value: 80/min).

Finally, ranking third were the patients with a sudden hearing loss and vestibular concurrent reaction, who were subjected to an equilibrium test within the first weeks following the sudden hearing loss. The maximum value averaged 50 beats/min for the positional nystagmus.

By means of the *quantified positional test* the severe clinical picture of vestibular neuronitis (fig. 1–6) or the stage of irritation of Menière's disease (fig. 12–15) with pronounced vertigo can also be readily demonstrated. In the case of sudden hearing loss, the vestibular component is slighter both subjectively and objectively. Here, the sensorineural hearing loss capable of being objectified frequently combined with tinnitus stands more to the fore than vertigo.

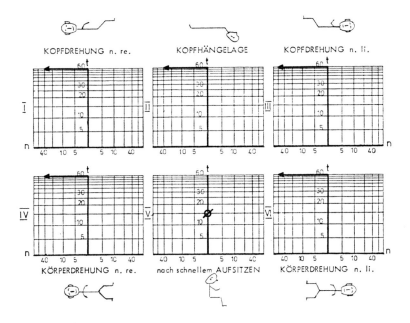

KOPFDREHUNG n. re.   KOPFHÄNGELAGE   KOPFDREHUNG n. li.

KÖRPERDREHUNG n. re.   nach schnellem AUFSITZEN   KÖRPERDREHUNG n. li.

Fig. 1. The positiogram of a 47-year-old patient 3 days after the onset of vestibular neuronitis on the left side in the acute stage. The test revealed a direction-fixed positional nystagmus beating to the right with the maximum nystagmic intensity in the case of head and body rotation to the left (80 beats/min).

Details

3 cases were chosen to illustrate the advantage of a quantitative recording of the results during the positional test by means of the positiogram [*Haid*, 1977]. The first case presented with vestibular neuronitis with remission of the pathological findings. In the second case this diagnosis had to be reviewed because of the progredient pathological results during the positional test when a neurinoma of the acoustic nerve was found to exist. The last case is to illustrate the severity of vertigo during the stage of irritation in Menière's disease.

The first patient was suddenly attacked in the morning by an intense permanent rotatory vertigo in bad as is typical of a vestibular neuronitis. The ENT status and the audiogram were normal. During the first vestibular function test, a direction-fixed spontaneous nystagmus, a direction-fixed positional nystagmus (fig. 1), as shown by the positiogram, a disturbance of the vestibular spinal reactions and a caloric inexcitability on the left side (fig. 2) were the prominent features of this acute disease. In the follow-up check 6 weeks later, there was only a slight positioning nystag-

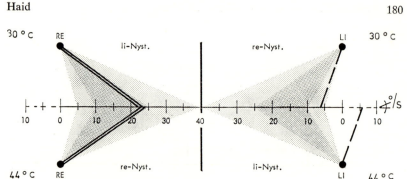

Fig. 2. The SPV calorigram of the same patient as in figure 1 shows an un-excitability on the left side when the vestibular neuronitis is in the acute stage. Normal excitability on the right side with an angular speed of about 23°/s during the two irrigations. The two dotted lines on the left side represent a spontaneous nystagmus with an angular speed of 6°/s. The bright and dark hatched fields give the values for normals. Values ranging between 5 and 40°/s during the culmination time of 30 s can be found in normals to almost 90 %.

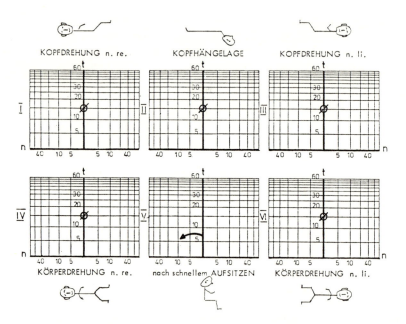

Fig. 3. About 7 weeks after the onset of the disease, the positiogram of the same patient as in figure 1 and 2 shows only a positioning nystagmus with a beat rate of 8 and a duration of 8 s after the rapid sitting-up.

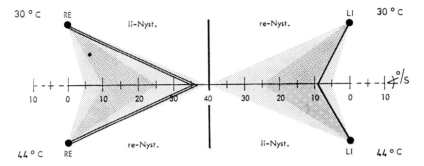

30 ° C   RE   li-Nyst.   re-Nyst.   Ll   30 ° C

re-Nyst.   li-Nyst.
44 ° C   RE   Ll   44 ° C

Fig. 4. The same patient as in figures 1–3. Excitability on the affected side, but to a clearly lesser extent than on the right side.

mus (fig. 3) as well as a subexcitability on the left side (fig. 4). During the last follow-up 2 months later, it was no longer possible to identify a pathological vestibular disorder (fig. 5, 6). The subjective complaints readily correlated to the objective ones.

The second patient was also suddenly seized by the typical permanent rotatory vertigo with vomiting as a significant feature of vestibular neuronitis. It is to be mentioned that some years before the patient suffered a sudden hearing loss in his left ear. The audiogram identified the sensorineural deafness also on the left side.

The Stenvers projections revealed no side difference of the internal auditory canals. During the first vestibular function test, the typical peripheral vestibular signs of vestibular neuronitis were identified, i. e. spontaneous nystagmus, the combination of a positional and positioning nystagmus (fig. 7), a disorder of the vestibular spinal reactions as well as a caloric inexcitability on the left side (fig. 8). During the follow-up some months later, a progressively pathological picture was obtained by way of the quantitative assessment of the results in the positiogram (fig. 9). The caloric reaction on the left side remained unchanged (fig. 10). Nowhere was a change seen in the audiogram. We carried out cisterno-meatography because an acoustic neurinoma was suspected on account of the progressively pathological results obtained from the positional test. This neuroradiological examination led to the detection of a tumour which was removed by otomicrosurgery. Histologically, this growth was found to be a neurinoma. During the last follow-up check 1.5 years after surgery, the pathological results originating from the positional test had clearly regressed (fig. 11). They are typical of the stage of the central compensation as a consequence of tumour extirpation with severence of the vestibular nerve.

The female patient of the last case presented with the typical history of Menière's disease (rotatory vertigo, unilateral tinnitus and hearing loss) when she appeared for the first vestibular function test. The audiogram revealed the typical omnifrequent and sensorineural hearing disorder in the left ear. During the positional test, an intense direction-fixed positional nystagmus beating toward the diseased

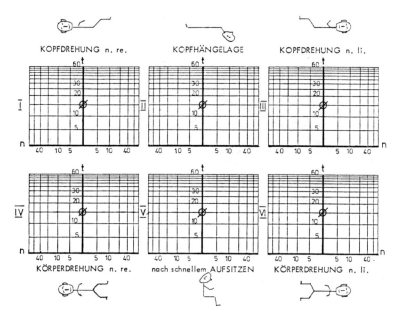

Fig. 5. Same patient as in figures 1–4, almost 4 months after the onset of vestibular neuronitis. During the positional test, no nystagmus was detected in the patient.

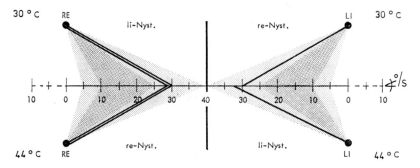

Fig. 6. The same patient as in figures 1–5. Both labyrinths proved to be excitable again homolaterally.

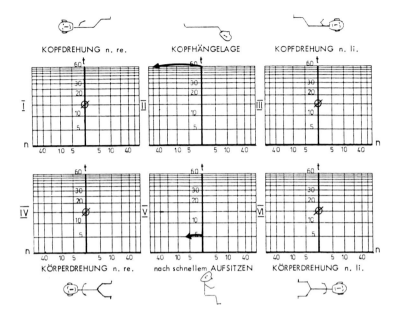

Fig. 7. The positiogram of a 60-year-old patient 5 days after the onset of an at first suspected vestibular neuronitis on the left side. The position II revealed a positional nystagmus to the right with a beat rate of 80 during 1 min and in position V a positioning nystagmus to the right with a beat rate of 5 and a duration of 5 s.

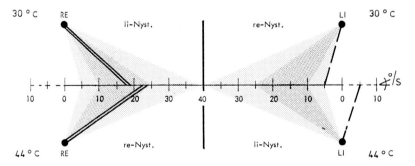

Fig. 8. The SPV calorigram of the same patient as in figure 7 shows an un-excitability on the left side.

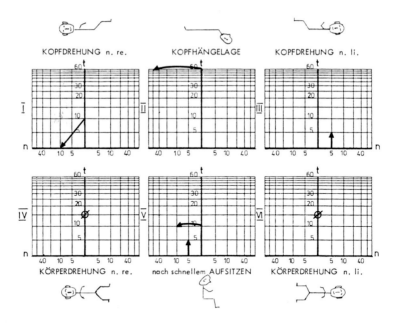

Fig. 9. In the positiogram of the same patient as in figures 7 and 8, the progress of the nystagmic intensity is discernible some months later. This led the examiner to suspect an acoustic neurinoma and this diagnosis was confirmed at surgery.

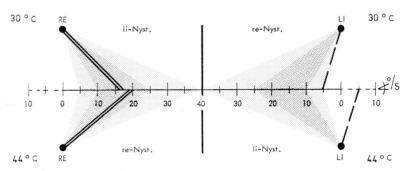

Fig. 10. The SPV calorigram of the same patient as in figures 7–9 shows the same result as in figure 8.

Fig. 12. The positiogram of a 69-year-old female patient shows during the irritation stage of Menière's disease an intense positional nystagmus beating toward the affected ear with the head and body turned to the same side.

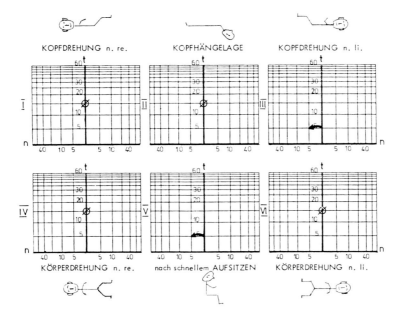

Fig. 11. In the positiogram of the same patient as in figures 7–9, a clear remission of the nystagmus intensity following excision of the neurinoma is discernible.
The patient is in the stage of central compensation.

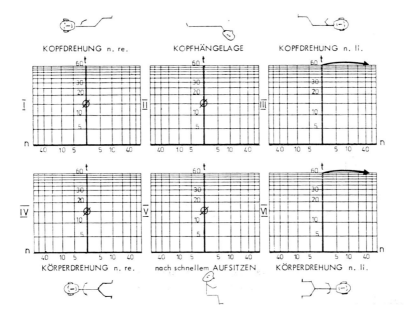

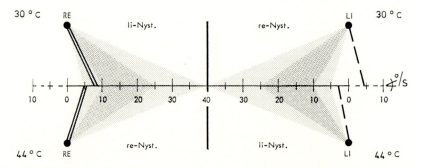

Fig. 13. The SPV calorigram of the same female patient as in figure 12 reveals unexcitability of the affected side during the irritation stage of Menière's disease.

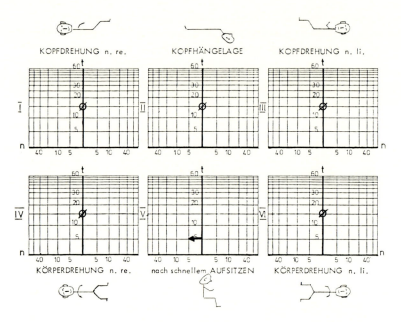

Fig. 14. During the interval stage of Menière's disease in the same female patient as in figures 12 and 13, only a slight positioning nystagmus was identified 6 months later.

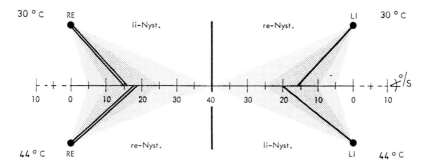

Fig. 15. During the interval stage of Menière's disease in the same female patients as in figures 12–14, the caloric stimulation resulted in homolateral excitability during the follow-up check 6 months later.

ear developed (fig. 12) as may often be typical of the stage of irritation. During the caloric test, the afflicted side proved inexcitable (fig. 13). On the occasion of the second equilibrium test 6 months later, a vestibular disturbance was objectivated. The audiogram manifested a clear improvement of the hearing ability. During the positional test, only a slight positioning nystagmus developed (fig. 14) as a pointer to the interval stage. The caloric test showed a homolateral and normal excitability (fig. 15). The fluctuating audiological and vestibular results are typical of Menière's disease.

Discussion

The vestibular function test is made up of numerous subtests, the positional test occupying the pride of place. The positional test can often be the only means of obtaining evidence of a vestibular disorder [*Gramowski* et al., 1973; *Stenger,* 1966] and this we can fully confirm. To us, it appears to be very important that the nystagmus beats occurring during the positional test can be quantified. This gives the possibility, especially in the case of many acutely occurring vestibular diseases, to read off the severity of the vertigo also by means of the positional test, e. g. during the acute stage of vestibular neuronitis (fig. 1–6) or during the stage of irritation of Menière's disease (fig. 12–15).

Quantification of the nystagmus by the use of the positional test allows above all the progress, constancy or remission to be identified during the follow-up checks (intra-individual comparison). The process of central compensation for a peripheral-vestibular disorder, e. g. following a permanent functional breakdown of a labyrinth or after neurectomy of the vestibular nerve on one side can also be readily demonstrated through quantification.

Progressive or protracted constant pathological results obtained during the follow-up checks after the positional test, as is demonstrated in figures 7–11 of a patient who was at first suspected of vestibular neuronitis, but also after a sudden hearing loss with vestibular involvement, definitely constitute an indication for the neuroradiological appraisal, allowing an acoustic neurinoma to be ruled out as cause.

However, in the early diagnosis of an acoustic neurinoma, the positional test proved to be one of the most sensitive examinations in addition to tone threshold determination and was found to be still superior to the caloric test. 95 % of all 78 patients in whom an acoustic neurinoma was diagnosed yielded pathological findings during the positional test [*Haid*, 1980]. 11 of the 12 patients affected by a small neurinoma gave pathological results during the positional test (92 %), while during the caloric test only 9 of the 12 patients showed subexcitability or non-excitability (75 % on the tumour side).

The positional test releases a minimum stimulus and the caloric stimulation triggers a maximum stimulus. The positional test may stimulate all three semicircular canals, the otolith system and higher vestibular centres, whereas the caloric test excites the superior vestibular nerve exclusively. Most neurinomas start growing on the inferior vestibular nerve [*Henschen*, 1916; *Nager*, 1969; *Ylikoski* et al., 1979]. This may furnish an explanation for the higher sensitivity of the positional test as compared with the caloric test.

Summary

With the aid of the positiogram, a quantitative recording of the positional test, it was possible to study the intensity of nystagmus in three acute vestibular diseases (vestibular neuronitis, Menière's disease and sudden hearing loss with vestibular involvement). Patients with vestibular neuronitis in the acute stage showed a higher intensity of nystagmus during the positional test than patients passing through the irritation stage of Menière's disease. Especially patients suffering from vestibular neuronitis complained also subjectively about such an intense vertigo that they were seized by a feeling of annihilation at the onset of the disease.

Quantification of the nystagmus during the positional test allows the progress, constancy or remission of the findings to be identified during the follow-up checks. Protracted constant pathological results obtained during the positional test, or when they become progressive, constitute in the light of our experience an indication for neuroradiological clarification in order to rule out an acoustic neurinoma as a cause (especially when a unilateral sensorineural hearing impairment exists).

References

Gramowski, K. H.; Unger, E.; Weinaug, P.: Zur Ursache von Lage- und Lagerungs-nystagmus. Mschr. Ohrenheilk. *7:* 285 (1973).

Haid, T.: Das Positiogramm, eine Aufzeichnungsmethode der Lageprüfung in der Neurootologie. Lar. Rhinol. *56:* 1037 (1977).

Haid, T.: Früherkennung des Akustikusneurinoms durch quantitative Neurootologie und radiologische Feindiagnostik; Habilitationsschrift (1980).

Henschen, F.: Zur Histologie und Pathogenese der Kleinhirnbrückenwinkeltumoren. Arch. Psychiat. Nervkrankh. *56:* 20 (1916).

Kornhuber, H. H.: Physiologie und Klinik des zentral-vestibulären Systems; in Berendes, Link, Zöllner, Lehrbuch Hals-Nasen-Ohrenheilkunde, vol. III/3, p. 2150 (Thieme, Stuttgart 1966).

Nager, G. T.: Acoustic neurinomas: pathology and differential diagnosis. Archs Otolar. *89:* 252 (1969).

Stenger, H. H.: Lagenystagmus-Lagerungsnystagmus; in Berendes, Link, Zöllner, Lehrbuch Hals-Nasen-Ohrenheilkunde, vol. III/3 (Thieme, Stuttgart 1966).

Ylikoski, J.; Collan, Y.; Palva, T.: Functional and histological findings in acoustic neuroma. ORL *41:* 33 (1979).

Dr. T. Haid, PD, ENT Clinic, University of Erlangen-Nürnberg, Waldstrasse 1, D-8520 Erlangen (FRG)

Adv. Oto-Rhino-Laryng., vol. 27, pp. 190–197 (Karger, Basel 1981)

Vertigo Originating from Inflammation of the Paranasal Sinuses (the So-Called Sinugenic Vertigo)

T. Haid

ENT Clinic at the University of Erlangen-Nuremberg, Erlangen, FRG

Introduction

The long established rhinological experience that patients with acute or chronic sinusitis feel giddy fairly frequently has led to this being defined as the so-called 'sinugenic vertigo' [*Appaix and Striglioni,* 1959; *Kissel* et al., 1960; *Terracol,* 1961]. In this paper, we wish to demonstrate that if, among other things, a patient complains of vertigo, a routine X-ray examination of the paranasal sinuses can make an important contribution to the diagnosis and therapy of vertigo.

Material

Between January 1974 and December 1979, 'sinugenic vertigo' was diagnosed in 30 patients out of a total of 4,112, which amounts to nearly 1 %. In addition to the audiological and neurotological investigations, including ENT status, routine X-ray examinations were made of paranasal sinuses and petrous bones.

Results

Of the 30 patients presenting with a pathological diagnosis of the sinuses, 12 were affected on both sides, 10 on the right side only and 8 on the left. Notable here was that in 28 patients the pathological X-ray finding was localized in the maxillary sinus. 2 of the patients suffered from

Table I. Types of vertigo in 30 patients with the diagnosis 'sinugenic vertigo'

Rotatory vertigo	4
Rotatory vertigo following abrupt movements of the body	12
Staggering vertigo	7
Feeling of falling	3
Feeling of floating	0
Feeling of unsteadiness	1
Feeling of numbness	3

a unilateral pansinusitis. Another interesting result was that, in seven instances, the survey view revealed a hyperplastic marginal cushion, in 12 cases, massive opacification, in 6 cases, diffuse clouding and, in three instances, round shadows, as an indication of cysts or polyps. Two pathological findings were weakly positive and 2 patients had undergone operation on the maxillary sinuses.

In 15 cases, the pathologic X-ray findings were confirmed during surgery, antroscopy, X-ray follow-up or by puncture of the maxillary sinuses. Unfortunately, the other half of the patients did not appear for the proposed control checks. Consequently, nothing is known about the further course of the disease in those 15 persons. In most cases, the patients described the vertigo as a systematic phenomenon. Roughly half of them complained of an acutely occurring rotatory vertigo which lasted only for a few seconds, however. It occurred in particular after sudden movements of the body. Only 1 patient reported an additional tinnitus and hearing loss (table I).

During the equilibrium test (table II), spontaneous nystagmus was found in only 4 patients. There were 13 instances of head-shaking nystagmus and 8 patients clearly showed lymphokinetic episodes, i. e. to-and-from eye movements (like 'jelly').

The positional test revealed that only 13 patients were afflicted by nystagmus (43 %), mainly in the form of a direction-changing positioning nystagmus. The direction of the nystagmus showed no correlation to the side in which the maxillary sinus was inflamed. 40 % of the patients were found to suffer from disturbed vestibular spinal reactions. The same percentage of patients presented a pathological caloric reaction, usually in the form of a one-sided caloric deficit.

Table II. Results of the vestibular function test in patients with 'sinugenic vertigo'

		Path./Number	%
1	Spontaneous nystagmus	4/30	13
2	Head-shaking nystagmus with lymphokinetic episodes	13/30	43
3	Nystagmus in the positional test	13/30	43
	(a) Positional nystagmus	2/30	
	(b) Combination of positional and positioning nystagmus	0/30	
	(c) Positioning nystagmus	11/30	
4	Pathologic vestibular-spinal reactions	12/30	40
5	Pathological caloric reaction	12/30	40

Table III. Results of the 'sinugenic vertigo' after treatment of the paranasal sinuses (follow-up checks only in 15 out of 30 patients)

No more vertigo	10
Remission of vertigo	4
No abatement of vertigo	1
Increase in vertigo	0

In addition to the pathological findings in the maxillary sinus, the vestibular function test frequently revealed an isolated positioning nystagmus or an isolated head-shaking nystagmus combined with disordered vestibular spinal reactions or merely a pathological caloric reaction.

Of the remaining 15 patients who returned for the proposed control investigations, 8, with sinugenic vertigo, were treated conservatively and 7 underwent surgery. It is pleasing to report that 10 patients were relieved of the distressing vertigo as a result of the treatment. The vertigo usually disappeared abruptly following surgery of the maxillary sinus. 4 patients admitted that dizziness had abated and only one claimed that the condition had not improved (table III).

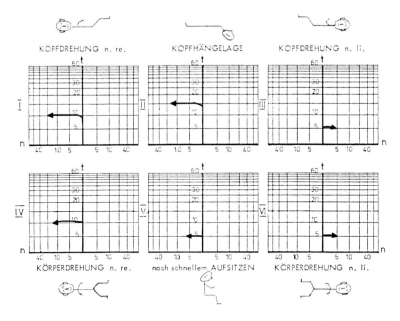

Fig. 1. The positiogram of a 52-year-old patient with 'sinugenic vertigo' reveals a partly direction-changing nystagmus in all 6 positions. The beat rate varies between 5 and 20 and the nystagmus duration ranges from 5 to 15 s.

Individual Case Details

Briefly, this is the case of a patient with sinugenic vertigo in which the X-ray examination and the vestibular function test as well as the surgical result led to the diagnosis. The patient, 52 years old, had complained for many years of an acute intermittent feeling of falling to the right side and of an uncertain gait brought about by sudden body movements. The equilibrium test revealed spontaneous nystagmus, intense positioning nystagmus (fig. 1) and considerable disturbance of the vestibular spinal reaction (tendency to fall). During the caloric test, insensitivity was found on the left side (fig. 2).

The X-ray examination showed both maxillary sinuses as opacified areas (fig. 3). After surgical treatment of the sinuses from which pus under pressure was removed, the condition of the vertigo patient improved abruptly. Only 1 day after the surgical treatment, this improvement could be objectively observed in the vestibular function test. During the positional test, only a slight positioning nystagmus was detected (fig. 4). During the demonstration of the vestibular spinal reactions, the patient no longer experienced a tendency to fall. The caloric test, however, still revealed an insensitiveness on the left side, but in the frequency calorigram (fig. 5), a remission of the spontaneous nystagmus was recognized.

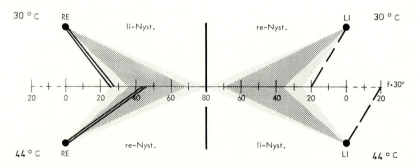

Fig. 2. The frequency calorigram of the same patient as in figure 1 shows insensitiveness on the left side. The dotted straight lines on the same side represent the spontaneous nystagmus to the right side with a rate of 20 beats during 30 s.

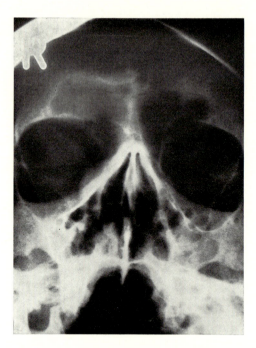

Fig. 3. The radiograph of the same patient as in figures 1 and 2 shows an opacification of both maxillary sinuses as indication to an inflammation.

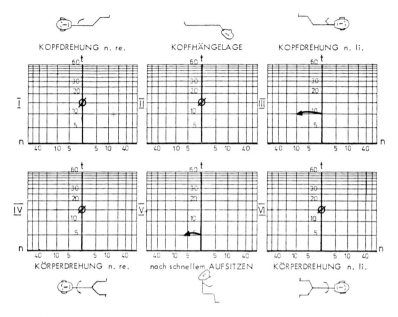

Fig. 4. In the positiogram of the same patient as in figures 1–3, a clear remission of the nystagmic intensity could be identified 1 day after the surgical sanitation of the paranasal sinuses.

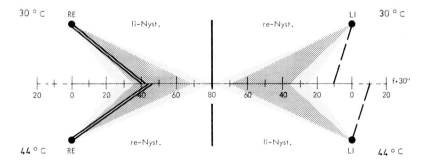

Fig. 5. The caloric test of the same patient as in figures 1–4 failed again to trigger a reaction in the left ear. However, it can be seen that the intensity of the spontaneous nystagmus has abated even on the first day after surgery (dotted lines).

Discussion

A striking fact was that 10 out of 15 patients were relieved of vertigo after maxillary sinus therapy had been carried out and that the vertigo diminished in 4 patients. In a former group of patients with acute labyrinthopathy of the Menière type, we were also able to observe a causal relationship between a maxillary sinus disease and labyrinthine irritation [*Haid*, 1975]: In several cases, the patients experienced an amelioration of the otic complaints in addition to the elimination of vertiginous attacks following surgery on the maxillary sinuses. The elimination of vertigo was also observed objectively in the vestibular function test. It was demonstrated that patients presenting a Menière-type pathological picture, as opposed to two comparative groups (patients with otosclerosis, as well as others with vestibular neuronitis or sudden hearing loss) furnished a considerably higher percentage of pathological X-ray findings for the sinuses (47 against 15 and 20 %, respectively).

To what extent a causal relationship exists between sinusitis and labyrinthine irritation has not yet been fully clarified and this should be the object of further investigations. It can at present only be speculated that interrelations exist between pathological trigeminus reflexes, via the sphenopalatine ganglion, triggered off by a maxillary sinusitis and a reflectory labyrinthine irritation resulting in vertigo. No evidence at all has been produced even for the assumption, of a serious labyrinthitis, but the correlation is of practical importance for the otologist because rhinological therapy may assist to a high degree in reducing or even in eliminating the vertigo. We therefore suggest that in the course of a neurotological investigation, the paranasal sinuses of a patient should always be X-rayed as a matter of routine since sinusitis may possibly be regarded as triggering systematic vertigo.

Summary

A routine X-ray examination of the sinuses of a patient complaining of regular bouts of dizziness may provide diagnostic information about a so-called sinugenic vertigo. In addition to the pathological X-ray findings in the maxillary sinuses, the patients presented either a positioning nystagmus or a head-shaking nystagmus, with disturbed vestibular spinal reaction as a pathological vestibular condition.

Out of 15 patients in whom a sinusitis-induced (sinugenic) dizziness was diagnosed and who appeared regularly for the control checks, 14 patients said that they were relieved of the dizziness as a result of sinus therapy, often immediately afterwards. Interrelationships possibly exist between pathological trigeminus reflexes via the sphenopalatine ganglion brought about by maxillary sinusitis and a reflectory labyrinthine irritation, triggering the vertigo.

References

Appaix, A.; Striglioni: Le vertige d'origine sinusienne, à propos d'un cas. Revue Oto-Neuro-Opthtal. *31:* 483 (1959).

Haid, T.: Die Häufigkeit von Kieferhöhlenentzündungen bei Patienten mit Menière-Anfällen. Archs Oto-Rhino-Lar. *210:* 354 (1975).

Kissel, P.; Grimaud, R.; Tenenbaum, P.: Vertiges d'origine sinusienne. Revue Oto-Neuro-Ophtal. *32:* 170 (1960).

Terracol, J.: Les vertiges d'origine sinusienne. Confinia neurol. *21:* 223 (1961).

Dr. T. Haid, PD, ENT Clinic, University of Erlangen-Nürnberg, Waldstrasse 1, D-8520 Erlangen (FRG)